TABLE OF CONTENTS

This handbook summarizes current agricultural engineering recommendations for beef producers. It deals with the design and operation of buildings and the equipment necessary for an efficient beef operation.

Up-to-date housing and facilities are important in beef production. They save labor; protect animals, feed, and equipment; and aid effective management.

When planning facilities, collect ideas from publications, farm visits, experienced cattle producers, animal scientists, and agricultural engineers. When visiting other operations, find out what changes the owner would make if rebuilding. Also ask about equipment used.

A list of selected references, additional plans, and an index are included at the end of the book.

1. PLANNING DATA SUMMARY

This chapter is a concise summary of much of the planning data presented throughout the handbook. These tables are located at the front of the book for easy reference. See the specific page references in Table 1-1 for more detailed discussion.

Table 1-1. Planning data.

| | | Feeder cattle | Bred | | | | |
	Calves 400-800 lb	Finishing 800-1,200 lb	heifers 800 lb	Cows 1,000 lb	Cows 1,300 lb	Bulls 1,500 lb	Page no.
Lot space			-- ft²/animal --				
Unpaved lot with mound (includes mound space)	150-300	250-500	250-500	300-500	300-500	1,200	5.1,5.7,6.1
Mound space	20-25	30-35	30-35	40-45	40-45	50-60	6.5
Unpaved lot without mound	300-600	400-800	400-800	500-800	500-800	1,500	
Paved lot	40-50	50-60	50-60	60-75	60-75	—	
Barn space							
Barn with lot	15-20	20-25	20-25	20-25	25-30	40	5.2,5.7,6.2
Barn without lot	20-25	30-35	30-35	35-40	40-50	45-50	5.2,5.7,6.2
Enclosed barn slotted floor	17-20 ft²/1,000 lb		-- not recommended --				6.6
Feeder space			-- in./animal --				8.1,8.7
Once-a-day feeding	18-22	22-26	22-26	24-30	26-30	30-36	
Twice-a-day feeding	9-11	11-13	11-13	12-15	12-15	—	
Self fed grain	3-4	4-6	4-6	5-6	5-6	—	
Self fed roughage	9-10	10-11	11-12	12-13	13-14	—	
Approximate feed requirements to estimate storage			-- lb/animal per day --				
Hay	4.8	3.5	3.5	25-30	25-30		
Haylage	9.6	7.0	7.0				
Corn silage	19.2	13.7	13.7				
Grain	6.5-9.5	5.5-7.5	5.5-7.5				
40% supplement	1.3	0.9	0.9				
Water[a]							9.1
Animals/cup or bowl — Lot	25	20	20	20	18	16	
Pasture	18	—	15	15	14	10	
Animals/ft of accessible tank perimeter — Lot	16	16	16	16	16	9	
Pasture	10	10	10	10	10	7	
Gal/hd-day — Hot weather	8-15	15-22	15	18	25	27	
Cold weather	4-7	8-11	7	9	13	14	
Ventilation — Cold barns	Provide ridge openings, eave inlets, and adjustable wall openings located low on sidewalls in the animal zone.						4.1-4.4
Warm barns	-- ft³/min per head --						
	15-100	20-130	30-180	50-470	50-470		4.4,4.5
Manure production[b] lb/hd-day	24-48	48-72	48	60	78	90	11.1
ft³/hd-day	0.4-0.8	0.8-1.2	0.8	1.0	1.3	1.5	

[a]Size system to provide full day consumption in a 4-hr period in hot weather.
[b]Total storage volume can be 25%-50% higher because of wasted/spilled feed and water.

Table 1-2. Bunk design.

Throat height (maximum)	
Calves (400-800 lb)	18″
Heifers/finishing (800-1,200 lb)	20″
Mature cows/bulls	24″
Bunk depth (maximum)	
Calves	8″
Heifers/finishing	12″
Mature cows	18″
Bunk width	
Eat from both sides	
Calves	36″
Heifers/finishing	48″-60″
Mature cows	48″-60″
Eat from one side	18″ bottom width
Mechanical feeder	Add 6″-12″ up to 60″ width
Step along bunk	
Height	6″-8″
Width	12″-16″
Bunk apron	
Slope	¾″-1″/ft
Width	10′-12′ (minimum)
Neck rails	
⅜″ tightly woven cable,	
2″ pipe, or 2x6 plank	16″-24″ opening

Table 1-3. Conversions.

Multiply to the right: acres × 43,560 = ft².
Divide to the left: ft² ÷ 43,560 = acres.

Unit	Times	Equals
Acres	43,560	ft²
	4,840	yd²
	160	square rods
	1/640	square mile
Acre-ft	325,851	gallons
	43,560	ft³
Acre-in	27,154	gallons
	3,630	ft³
Acre-in/hr	453	gpm
	1	cfs (approximate)
Bushels	1.25	ft³
	2.5	ft³ ear corn
ft³	7.48	gallons
	1728	in³
	62.4	lb water
	0.4	bu ear corn
	0.8	bu grain
cfs	448.8	gpm
	646,272	gal/day
Cubic yard	27	ft³
concrete	81	ft² of 4″ floor
concrete	54	ft² of 6″ floor
Gallons	231	in³
	0.134	ft³
	8.35	lb water
Miles	5,280	ft
	1,760	yd
	320	rods
Pressure, psi	2.31	ft of water head
Rods	16.5	ft
	5.5	yd

Table 1-4. Floor and lot slopes.

Handling facilities	⅛″-¼″/ft
Lots	
Paved	⅛″/ft minimum
Earth	½″-¾″/ft
Mound sideslope	1′/5′
Bunk apron	¾″-1″/ft nearly self-cleaning

Table 1-5. SI Conversions.

Multiply to the right: in. × 2.54 = cm
Divide to the left: cm ÷ 2.54 = inches

Unit	Times	Equals
Length		
inches	2.540	cm
feet	0.3048	m
yards	0.9144	m
miles	1.609	km
Area		
in²	6.451	cm²
ft²	0.09290	m²
yd²	0.8361	m²
acres	0.4047	hectares
mile²	2.590	km²
Volume		
in³	16.39	cm³
in³	0.01639	liters (L)
ft³	0.02832	m³
cubic yards	0.7646	m³
bushel	0.03524	m³
quarts (liquid)	0.9464	liters (L)
gallons (liquid)	3.785	liters (L)
Mass		
pounds	0.4536	kg
ton (2,000 lb)	907.2	kg
ton (2,000 lb)	0.9072	tonne (t)
Velocity		
ft/sec	0.3048	m/s
miles/hour (mph)	1.609	km/h
Flowrate		
cfm	0.0004719	m³/s
gpm	0.00006309	m³/s

Temperature
Fahrenheit (F); Celsius (C)
$C = (F - 32) \div 1.8$
$F = (C \times 1.8) + 32$

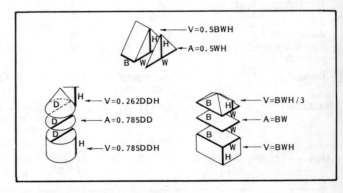

Fig 1-1. Areas and Volumes.

2. FARMSTEAD PLANNING

Planning objectives are expansion, improved performance, higher capacity, and better labor use. Many factors determine the best plan, and while some are common sense, overlooking one can cause a poorly planned farmstead. Collect ideas from publications, farm visits, county agents, experienced producers, scientists, and engineers. Plan on paper, where mistakes can be easily corrected. **It is less costly to correct a mistake during the planning stage than after construction begins.** Stake out the best arrangements on the site to see how they fit.

Consider the entire farmstead when planning a new or modified production system. Solving one problem may cause another. With proper planning and attention to details, a well organized, functional farmstead can result.

Site Selection

Space Between Major Activities

For planning assume your operation will double in size.

Distances between farm service buildings and feeding lots depend on management needs, operation size, and pollution hazards. Locate cattle feeding and handling facilities for easy access. In operations where managers spend much of their working day somewhere else, place facilities convenient to the farm service area. Locate high-labor facilities such as treatment and calving facilities within 300' and finishing and mature animal facilities at least 500' from the family living area, Fig 2-1. Large operations create more noise, odors, dust, and traffic, requiring greater separation distances. Larger operations with full time personnel can be farther from the living area where there is little movement back and forth to the rest of the farmstead.

Space Between Buildings

Provide space for new buildings, clearance between buildings, and expansion. Consider space for vehicle access and parking. Separate major buildings by at least 75' for access by fire fighting equipment. Space between buildings less than 75' is a compromise. Naturally ventilated buildings may require more than 75' between buildings for proper air movement. Determine required spacing—but no less than 75'—between naturally ventilated buildings using Eq 2-1. Use dimensions of the windward building based on winter winds to calculate minimum separation distance for winter. For summer ventilation, calculate the minimum separation distance based on south winds and nearby shelterbelts and buildings.

Eq 2-1.

$$SPC = 0.4 \times HGHT \text{ (ft)} \times \sqrt{LGTH \text{ (ft)}}$$

SPC = minimum separation distance between buildings
HGHT = total building height to the ridge
LGTH = total length of building

Example 2-1:

Calculate the proper winter and summer separation distance between two naturally ventilated buildings. Each has a 4:12 gable roof. The building to the north is 32'x72' with 10' sidewalls. The southern building is 40'x120' with 12' sidewalls. Assume that the prevailing winter winds are out of the northwest.

Solution:

Calculate the minimum winter separation distance using the dimensions of the 32'x72' building. Calculate the total building height (ridge height):

HGHT = Sidewall height (ft) + (Bldg width (ft) ÷ 2) × Roof slope

10' + (32' ÷ 2) × (4 ÷ 12) = 15.33'
Total length is 72'.
Minimum winter separation is:

$$SPC = 0.4 \times 15.33' \times \sqrt{72'} = 52'$$

Use a 75' separation distance for minimum winter separation because 52' is less than the 75' recommended for fire protection.

Assuming prevailing summer winds are out of the south, calculate the minimum summer separation distance based on the 40'x120' building.

Total building height (ridge height) is:

12' + (40' ÷ 2) × (4 ÷ 12) = 18.67'
Total length is 120'.
Minimum summer separation distance is:

$$SPC = 0.4 \times 18.67' \times \sqrt{120'} = 81.8'$$

Use an 82' separation distance between buildings for best summer ventilation.

Minimum year-round separation distance is 82' based on the summer separation distance.

Water

A year-round supply of water is essential for watering animals, sprinkling for dust control and cooling, veterinary use, fire protection, and possible manure dilution.

Design the watering system to meet maximum needs. Provide at least 18 gal/day in hot weather and about 9 gal/day in cold weather per 1,000 lb of animals. Peak water demand occurs shortly after feeding. Design a water system to provide 1 day storage and full day consumption in a 4-hr demand period. Where groundwater supplies are not adequate, use a rural community water system or surface sources such as a farm pond. The expense of surface sources may justify selecting a different site. For more information see the water and waterers chapter and MWPS-14, *Private Water Systems Handbook*.

Topography and Drainage

Topography affects drainage, building location, access routes, and prevailing wind directions. Do not

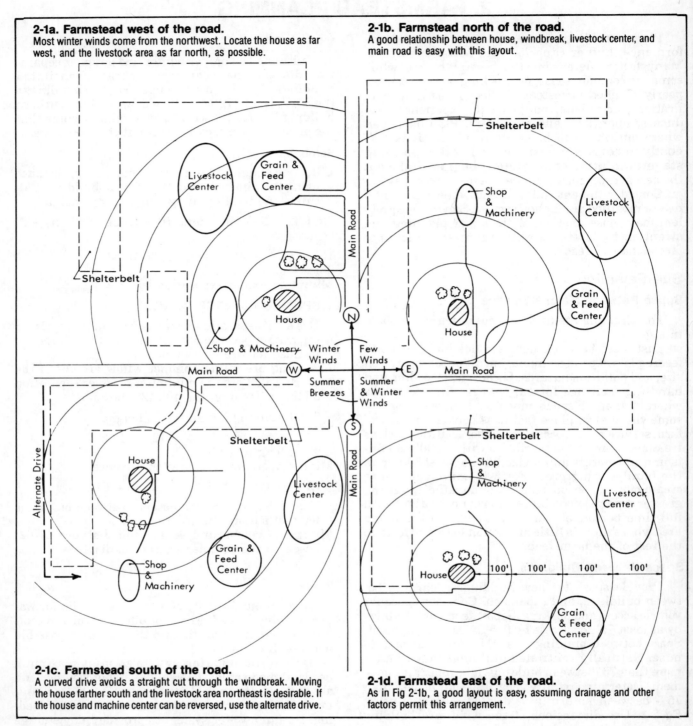

2-1a. Farmstead west of the road.
Most winter winds come from the northwest. Locate the house as far west, and the livestock area as far north, as possible.

2-1b. Farmstead north of the road.
A good relationship between house, windbreak, livestock center, and main road is easy with this layout.

2-1c. Farmstead south of the road.
A curved drive avoids a straight cut through the windbreak. Moving the house farther south and the livestock area northeast is desirable. If the house and machine center can be reversed, use the alternate drive.

2-1d. Farmstead east of the road.
As in Fig 2-1b, a good layout is easy, assuming drainage and other factors permit this arrangement.

Fig 2-1. Farmstead and main road relationships.
The direction of the farmstead from the main road affects farmstead layout. See MWPS-2, *Farmstead Planning Handbook,* for additional site planning information.

locate facilities on a flood plain, swale, low ground, peat soil, or very rocky soil.

Intercept and divert surface water from buildings, lots, and traffic areas. Construct diversion ditches or terraces across the slope to carry runoff away.

Slope lots 4%-6% away from buildings, feedbunks, and prevailing winter winds. A south or east slope is usually best. Lot slopes less than 4% require mounds to provide dry resting areas. Unpaved lots sloped over 8% can erode badly. Earth moving is inexpensive compared with facility costs.

Locate feedbunks along the high side of the lot or up and down the slope for best lot drainage. A north-south or northwest-southeast orientation is preferred to promote thawing and drying on each side of a feeder.

Subsurface drainage is needed around building foundations and below-ground storages to reduce frost heaving and hydraulic and soil pressures. Drainage from around below-ground manure storages must not pollute streams or groundwater. Avoid sites with springs and high water tables. Provide good drainage below all flat grain storage facilities to reduce moisture migration through the floor and grain.

Drives and Parking

Provide all-weather drives for year-round receiving and shipping of cattle and access by service persons, technicians, veterinarians, feed handling equipment, etc. Provide space for maneuvering vehicles, parking, and snow storage. Provide minimum drive widths of 14′ for vehicle traffic and 12′ for cattle

traffic. Include at least 7′ of additional clearance on each side for overhanging equipment and snow storage. Provide at least 3 parking spaces for visitors, in addition to employee and truck parking.

Service Area

Plan for a business and operations office, toilet room, parking, and access area. An office is needed for herd records, daily business activities, and protection for computer-based feeding and record systems. Provide a refrigerator for pharmaceuticals. Concrete masonry is a popular material for office buildings because of low first cost, low maintenance, durability, and fire safety. However, unless well insulated, concrete buildings are expensive to heat and cool. Possible office layouts are shown in Figs 2-3 and 2-4.

In the toilet room, provide at least a toilet and sink. Consider a shower, lockers, and a rest area for hired workers, Fig 2-5. Separate facilities for men and women may be required by local regulations.

Locate shipping and receiving facilities near the office for easier supervision.

Manure Storage and Handling

Locate, size, and construct storages for convenient filling and emptying and to keep out surface runoff. Plan for adequate storage capacity. Provide enough storage to spread manure when field, weather, and local regulations permit. Many states have minimum storage requirements.

Select a site with enough land nearby for spreading manure. State agencies often base minimum acreage on satisfying the nitrogen requirement of the growing crops. Manure handling methods that result in nitrogen loss, such as lagooning and aeration, decrease the amount of land required. Consider asking neighbors about manure spreading privileges.

Avoid steep slopes where manure runoff can cause water pollution, and avoid land adjacent to neighboring residences.

Grain-Feed Center

Locate a grain-feed center near the major feed users and for convenient vehicle access if hauling feed. Consider the following:

- **Leave room for growth.** More storage, faster drying, a feed processing center, or larger vehicles may be needed in the future. Plan for maximum mechanization.
- **Relationship to other buildings.** Place the feed-grain center at least 300′ from the home to reduce noise and dust and improve appearance. Locate facilities where prevailing winds are least likely to carry dust and other foreign material toward activity centers.
- **All-weather access roads.** Arrange facilities so transport vehicles can drive through without opening gates or backing up. Consider requirements for feed delivery and livestock transport vehicles.

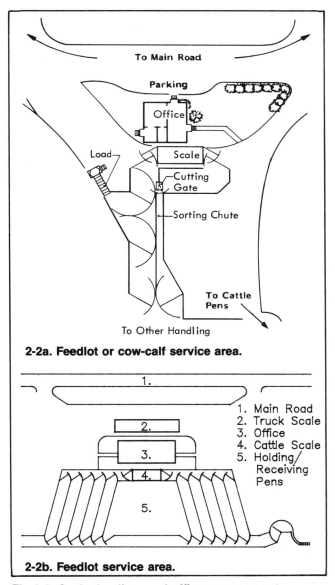

2-2a. Feedlot or cow-calf service area.

1. Main Road
2. Truck Scale
3. Office
4. Cattle Scale
5. Holding/ Receiving Pens

2-2b. Feedlot service area.

Fig 2-2. Scale, loading, and office arrangement.

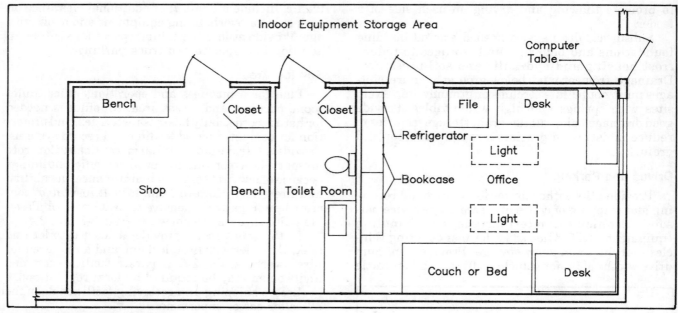

Fig 2-3. Office and shop layout.
Office and shop in the corner of a feeding or equipment storage building.

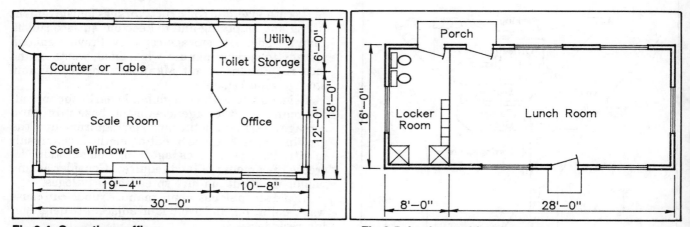

Fig 2-4. Operations office.
Office for a small feedlot. The scale room can be a hired help rest area.

Fig 2-5. Locker and lunch room.

- **Soil and drainage.** Test soil for strength to support heavy storages and equipment. Choose a well drained site. Provide drainage for dump pits and perimeter drains around footings. Polyethylene vapor barriers are essential beneath floors of grain storage.
- **Utilities.** Check for adequate single phase or 3-phase power and/or petroleum fuel before buying equipment. Many utilities limit single phase motors to 10 hp. A phase converter is an option in some cases. Phase converters are about 85% efficient, so higher operating costs may offset the benefits.

Electrical Power

Electrical power is needed for ventilating, heating, lighting, pumps, and motors. A 100 amp, 240 volt single phase service entrance is common. Several large motors may justify 3 phase power. A major need is feed handling and processing. Consider standby emergency power in the event of a power outage—especially for water pumping and heated waterers.

Security

Consider theft, vandalism, and fire safety. Limit visitor access to control disease and to reduce interference with farm work. If located on the same farmstead as the manager's residence, a single access road that is visible from the residence is best. If a second access is used for feed, manure, and animal transport vehicles, provide a gate and/or alarm system to control unauthorized traffic.

Facilities remote from the manager's residence pose the most problems. Provide only one access road—unauthorized persons are less apt to visit if there is no escape route should the manager return. If possible, make access roads at remote sites visible from a public road or neighboring residence.

Remodeling

When planning to expand, you may have to decide whether to remodel or abandon an existing building. Use a scale drawing of the overall layout. Carefully consider future as well as present needs. Evaluate these general factors:
- Compatibility to the final setup and future plans.
- Structural condition of building.
- Location of existing building.
- Cost of remodeling vs. a new building.

Remodeling is not always the cheaper route, especially when future needs are considered. If remodeling cost is more than ⅔ of the new building cost, a new building is usually best. Often, it is possible to use some materials from the existing building for the new building.

Plans, Specifications, and Contracts

Detailed documents help provide needed communication and understanding between owner, builder, and lender. **Plans** show all necessary dimensions and details for construction. **Specifications** support the plans; they describe the materials to be used, including size and quality, and often outline procedures for construction and quality of workmanship. The **contract** is an agreement between the builder and the owner; it includes price of construction, schedule of payments, guarantees, responsibilities, and starting and completion dates.

You have several options for preparing this material.

Be your own contractor. Draw a final plan, making sure dimensions are correct and construction details and materials are determined. Have plans checked by the appropriate regulatory agency when required. Determine total costs before beginning construction.

Hire a consulting engineer.

Use a design and construction firm. Some firms make working drawings and specifications and have standard contract forms.

Consulting Engineers

Services commonly offered by consulting engineers include:
- Direct personal service (technical advice, etc.).
- Preliminary investigations, feasibility studies, and economic comparison of alternatives.
- Planning studies.
- Design.
- Cost estimates.
- Engineering appraisals.
- Bid letting.
- Construction supervision and inspection.

Consulting engineers usually do a project in three phases: preliminary planning, engineering design, and construction monitoring. They may be retained to help with one or more of these phases. To select a consulting engineer, consider:
- Registration: to protect the public welfare, states certify and license engineers of proven competence. Practicing consulting engineers must be registered professional engineers in their state of residence, and qualified to obtain registration in other states where their services are required.
- Technical qualifications.
- Reputation with previous clients.
- Experience on similar projects.
- Availability for the project.

Wind and Snow Control

Windbreak fences and shelterbelts protect against wind and snow around buildings, along roads and drives, and in open lots. Windbreaks change wind direction but do not stop wind. Shelterbelts can reduce winds and trap a major part of heavy snows outside the farmstead. Use local experience to determine windbreak type, position, distance from buildings, and successful tree species. Take advantage of hills, trees, buildings, and haystacks for winter wind protection. Allow for summer air movement and drainage when locating windbreaks. Consider prevailing wind directions for reducing odors, snow drifting, insects, and noise, Fig 2-14.

Windbreak Fences

A solid fence provides better wind protection for short distances, but snow accumulates near the fence. Wind passing over a vertical barrier usually drops or swirls downward on the downwind side, losing energy and dropping snow. An 80% solid fence reduces wind speeds for a greater distance and spreads the snow out for faster melting, Fig 2-6.

Provide about 20% open space in a windbreak fence for effective wind and snow control, Table 2-1. Rough cut or dressed lumber 6"-10" wide works best for slotted windbreak fences. Plywood or metal roofing sheets are not as effective. Slot openings greater than about 2" allow too much wind through at one location. Several narrow slots (less than 2") allowing the same total amount of wind through, provide better wind and snow protection. A minimum height of 8' for solid fences and 10' for slotted fences is recommended for better wind control.

Horizontal or vertical slots in a slotted fence perform about the same. Leave a 4"-6" opening under the fence for better drainage, drying, and summer air movement. Close the opening below the fence with straw or snow in the winter to reduce drafts.

Attach boards on the cattle side of the fence. Install a horizontal rub rail if cattle have access to both sides. Use galvanized nails.

Snow fencing can provide localized wind and snow control. It is too porous to provide the best wind protection and is not durable enough for permanent use around cattle.

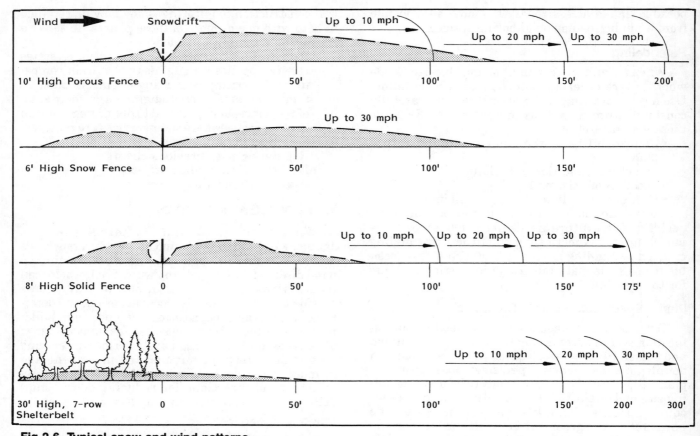

Fig 2-6. Typical snow and wind patterns.
With a 40 mph wind from the left, velocities are reduced to about those shown. For other wind speeds, reductions are proportional. Locate buildings away from the snow drift but within good wind protection.

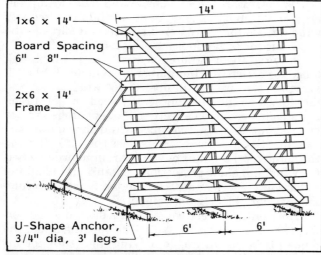

Fig 2-7. Moveable windbreak fence for pastures.
Several moveable fences end-to-end give local snow and wind protection.

Protecting Open-Front Buildings

Winter winds cause drifting and draft problems in open-front barns. Protect barns from wind and snow, Fig 2-9.

- Provide a continuous eave inlet on the closed side. Adjust the inlet to keep out blowing snow.

Table 2-1. Windbreak fence board spacing.
Recommended spacing between planks is 20%-25% of plank width. Plank width greater than 10″ is less effective. Openings over 2″ wide are not recommended as too much wind can pass through in one place.

| Board size | Slot width | |
	Rough cut lumber	Dressed lumber
	- - - - - in - - - - -	
1x4	⅞	¾
1x6	1⅜	1⅛
1x8	1¾	1⅝
1x10	2	2
1x12	2	2

- Provide a 40′ or more wide wind "passage" between barns, silos, or other structures.
- Install slotted windbreak fences at the ends to form a swirl chamber and reduce the wind in front of the building. A 16′ minimum offset to the side and rear of the front corner is recommended.
- Close part of the front wall at each end—up to ⅙ of the building length at each end.
- Install solid cross partitions 50′ apart in long buildings to reduce drafts.

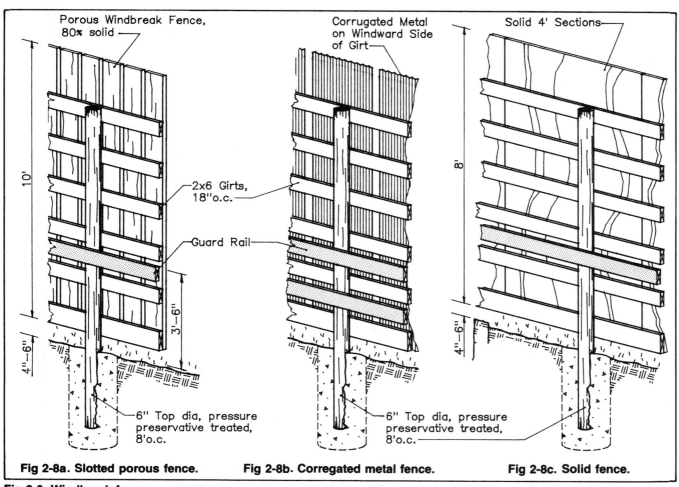

Fig 2-8a. Slotted porous fence. Fig 2-8b. Corregated metal fence. Fig 2-8c. Solid fence.

Fig 2-8. Windbreak fences.
Attach boards on the cattle side. If cattle have access to both sides, add guard rails. Splice framing planks on alternate posts.

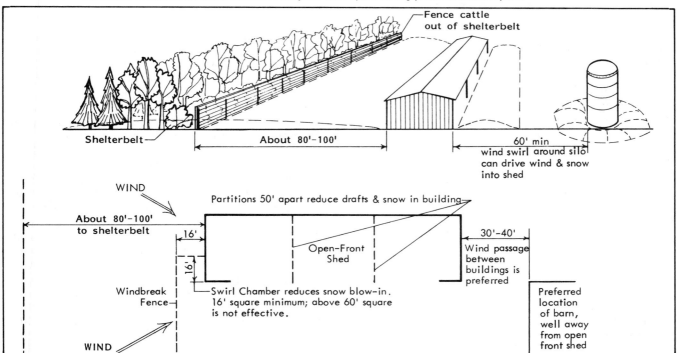

Fig 2-9. Wind and snow protection of open-front buildings.

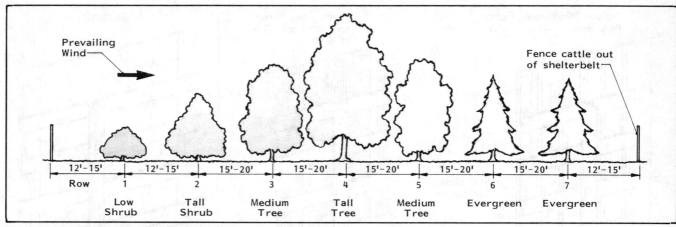

Fig 2-10. Typical shelterbelt tree planting.

Shelterbelts

Shelterbelts provide both snow and wind protection. In the Midwest, shelterbelts are usually on the north and west sides of the farmstead. Where winter winds from the south cause snow drifting into the farmstead, plant a 2-row hedge across the south. Shelterbelts reduce wind velocities 1 to 2 tree heights upwind and 10 to 20 tree heights downwind. The shelterbelt width can vary from 40'-150' (2 to 7 rows) depending on local conditions.

Plant multi-row shelterbelts of tall and short trees, evergreens, and shrubs for best wind and snow protection. Plant low, dense-growing shrubs in the outside rows to trip the snow as it enters the shelterbelt. Next should be the medium sized trees and then the tall growing trees in the center. Plant evergreens, such as spruces, cedars, and pines, on the inside of the shelterbelt closest to the buildings. Because snow whips around the ends, extend the shelterbelt 50' past the area to be protected. Allow 100' between the end of the shelterbelt and the road. Contact your local extension office for recommended tree species and spacing.

Space needed for trees depends on the shelterbelt density, height, width, and required snow catch area. Provide at least 95' between the inside of the shelterbelt and the nearest building.

For a large snow catch area, two shelterbelts 150'-300' apart can be used. The upwind shelterbelt can be 1 to 2 rows of trees at a field fence or on a terrace to reduce interference with other operations. A double shelterbelt usually reduces wind effects more than a single shelterbelt or windbreak fence.

Plan openings for lanes and roads to minimize the amount of snow and wind coming through. Plant trees and shrubs to avoid creating wind tunnels, Fig 2-11.

Shades

Shades provide comfort and cooling for cattle in pastures and open lots. Benefits from an effective

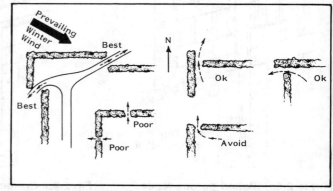

Fig 2-11. Traffic openings through tree shelterbelts.
Arrange openings to reduce snow and wind problems.

shade include improved feed intake for cattle and improved reproduction in cows.

If feed and water are provided under the shade, orient the long dimension east-west. Otherwise, orient the long dimension north-south to promote drying under the shade.

Build shades at least 12' high. Increase the height to improve air movement if shades are wider than 40'. Allow at least 50' between shades and trees, buildings, or other obstructions.

Provide at least 40 ft² of shade space per animal. For high rainfall areas, consider a paved area with a 1½%-2% slope to eliminate mud. Because cattle stand in the shadow of a shade, consider shadow migration throughout the day when planning the shade area.

The most effective shade materials are reflective surfaces such as white galvanized metal or aluminum. Painting the upper surface white and the lower surface black improves cooling. Shade cloth providing 30%-90% shade is available. Shade cloth with greater than 80% shade tends to hold water after rainfalls, causing structural problems.

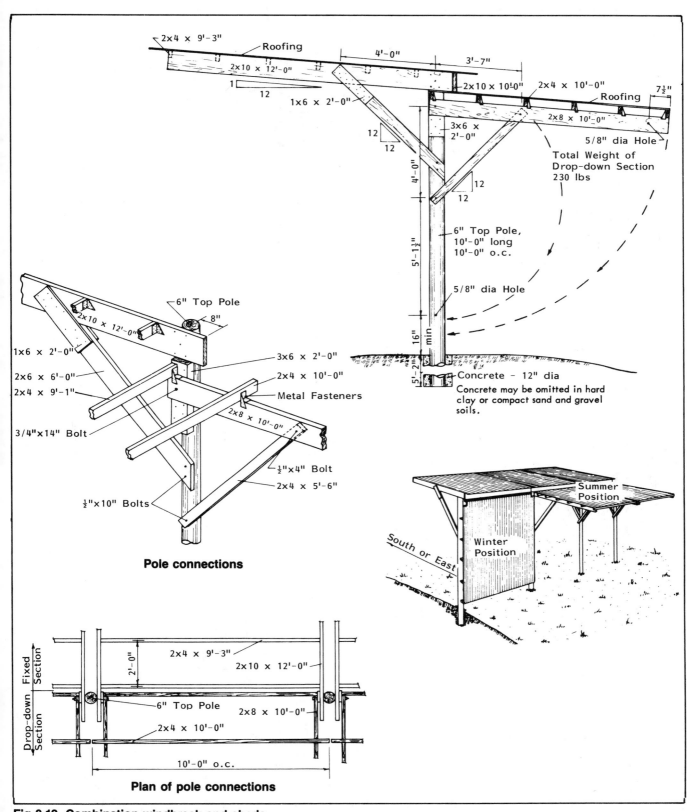

Pole connections

Plan of pole connections

Fig 2-12. Combination windbreak and shade.

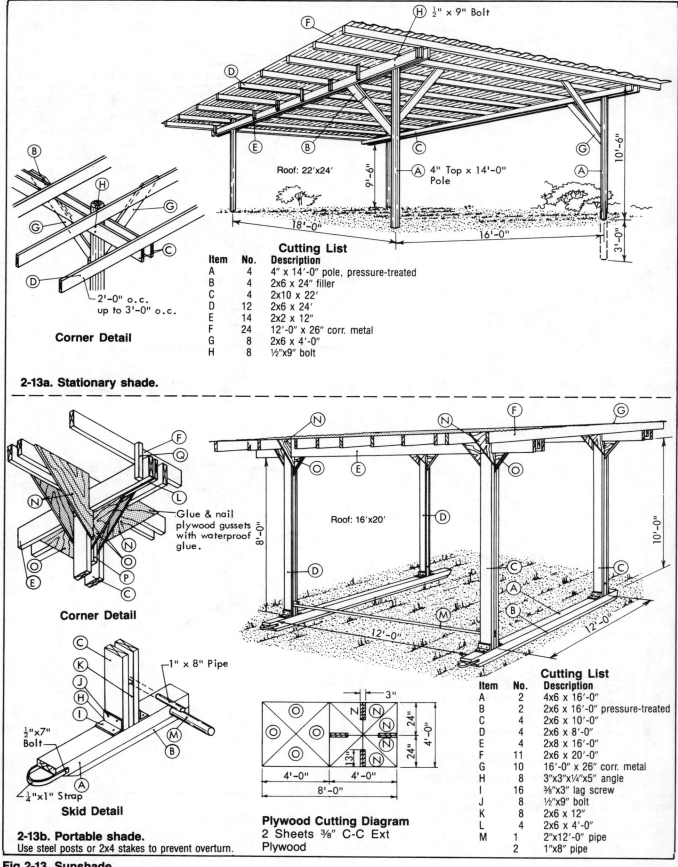

$\frac{1}{2}$" × 9" Bolt

F

D

H

E B C

G

A

Roof: 22'x24'

9'-6"

10'-6"

3'-0"

A 4" Top x 14'-0" Pole

18'-0"   16'-0"

Corner Detail

B

H

G

G

C

D

2'-0" o.c.
up to 3'-0" o.c.

Cutting List

Item	No.	Description
A	4	4" x 14'-0" pole, pressure-treated
B	4	2x6 x 24" filler
C	4	2x10 x 22'
D	12	2x6 x 24'
E	14	2x2 x 12'
F	24	12'-0" x 26" corr. metal
G	8	2x6 x 4'-0"
H	8	$\frac{1}{2}$"x9" bolt

2-13a. Stationary shade.

F
Q
L

N
O
E
P
C

Glue & nail
plywood gussets
with waterproof
glue.

Corner Detail

N N F G

O

E

D

Roof: 16'x20'

8'-0"

10'-0"

D

C
A
M
B

12'-0"   12'-0"

C

K
J
H
I

1" x 8" Pipe

$\frac{1}{2}$"x7"
Bolt

M
B

A

$\frac{1}{4}$"x1" Strap

Skid Detail

2-13b. Portable shade.
Use steel posts or 2x4 stakes to prevent overturn.

3"

O   N
N

24"

O   O

4'-0"

13"

24"

O   N
N

4'-0"   4'-0"

8'-0"

Plywood Cutting Diagram
2 Sheets $\frac{3}{8}$" C-C Ext
Plywood

Cutting List

Item	No.	Description
A	2	4x6 x 16'-0"
B	2	2x6 x 16'-0" pressure-treated
C	4	2x6 x 10'-0"
D	4	2x6 x 8'-0"
E	4	2x8 x 16'-0"
F	11	2x6 x 20'-0"
G	10	16'-0" x 26" corr. metal
H	8	3"x3"x$\frac{1}{4}$"x5" angle
I	16	$\frac{3}{8}$"x3" lag screw
J	8	$\frac{1}{2}$"x9" bolt
K	8	2x6 x 12"
L	4	2x6 x 4'-0"
M	1	2"x12'-0" pipe
	2	1"x8" pipe

Fig 2-13. Sunshade.

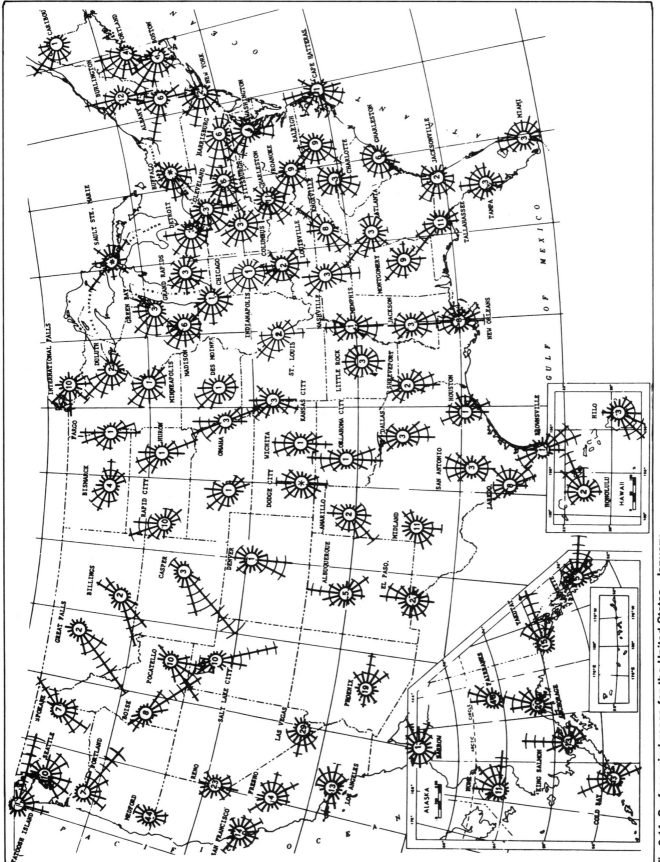

Fig 2-14. Surface wind roses for the United States, January.
U.S. Weather statistics from hourly observations, 1951-1960. Wind roses show % of time wind blew from the 16 compass points or was calm. * = less than 0.5% calm.

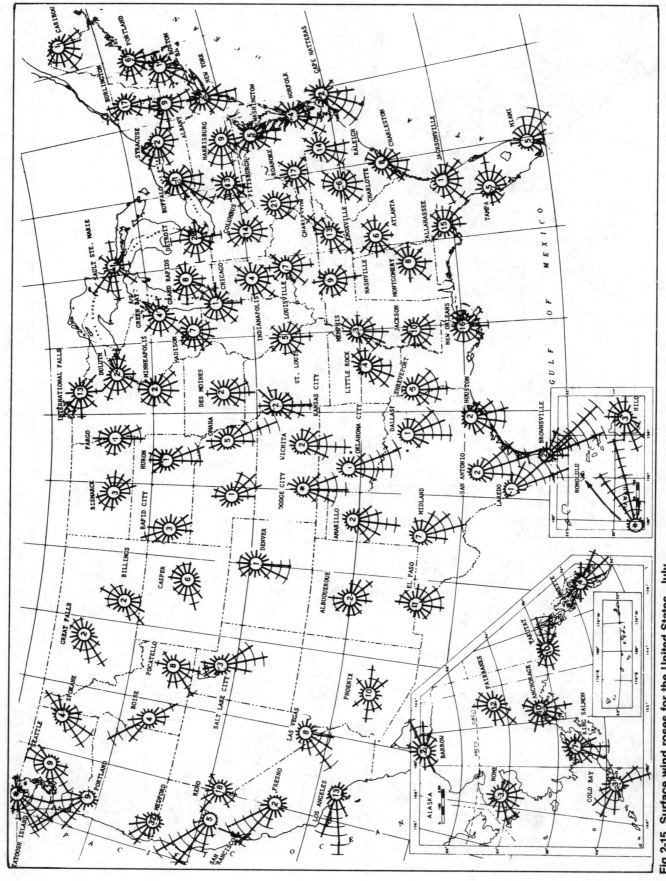

Fig 2-15. Surface wind roses for the United States, July.
U.S. Weather statistics from hourly observations, 1951-1960. Wind roses show % of time wind blew from the 16 compass points or was calm.

3. BUILDING CONSTRUCTION AND MATERIALS

Barns

Open-front or enclosed, one floor, clear-span barns are suitable for cattle housing. Clear-span barns are flexible for arranging pens and traffic lanes, are more adaptable, and make cleaning with mechanical equipment easier.

Uninsulated or single thickness walls are suitable for cold housing. Use pressure preservative treated wood or reinforced concrete at least 1' high along the wall to minimize wall deterioration.

Overhead hay and bedding storage requires post-beam construction for support. Although interior posts can interfere with layout flexibility, with careful planning, they can serve as pen partition supports. Also, hay and bedding stored overhead is good insulation. Barns with flat ceilings and lofts are usually more difficult to ventilate. They require mechanical ventilation, which adds to operation costs.

Barn style is determined by the roof shape. Common shapes are shed or monoslope, gable, offset gable, and saltbox. Pole barns with gable roof and clear-span wood trusses are the most common.

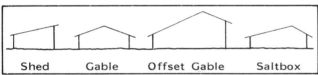

| Shed | Gable | Offset Gable | Saltbox |

Fig 3-1. Roof types.

A **shed or monoslope roof** is for open-front and enclosed buildings, attached lean-to additions, roof extensions, and small movable buildings.

A shed roof is low cost, provides good headroom, and is simple to build. Free-standing barns are easy to ventilate, but ventilating attached sheds and under roof extensions is often difficult. Many shed roofs have a low slope to keep the high side of the roof as low as possible. This can add to roof design load requirements and increase ventilation problems. Shed roofs are usually post and rafter construction but single slope truss designs are available.

A **gable roof** is common for both open-front and enclosed barns, Fig 3-2. It is moderate in cost (spans 24'-60') and fairly simple to construct. It is usually naturally ventilated through eave, sidewall, and ridge openings.

Pole barns with clear-span wood trusses are used extensively for cattle. Wood or steel rigid frames are also clear-span. Post and beam construction is popular in some areas, especially where native lumber is available.

The **offset gable roof** has equal roof slopes of different lengths. The ridge is off center in relation to the building width and the eave heights are different.

A **saltbox roof** has two different roof slopes of different length. The roof length is greater on the side with the lower slope. The ridge is off center with respect to the building width and the eave heights are equal.

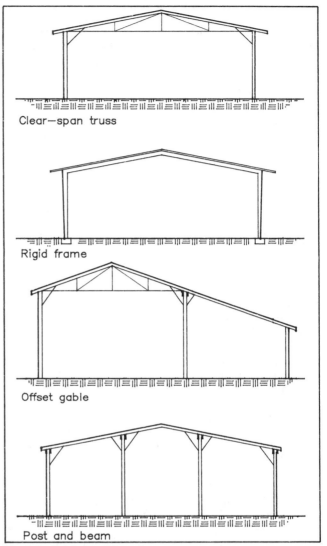

Clear-span truss

Rigid frame

Offset gable

Post and beam

Fig 3-2. Gable roof framing.

With offset gable and saltbox construction, framing commonly requires interior posts, although a truss or rigid frame gable roof on the high side of the barn may provide a clear-span for ⅔ or more of the barn width.

Lumber

Lumber is available as ungraded (native) or commercial lumber. Native lumber comes directly from saw mills or is cut from your own logs. Commercial lumber is usually graded and is available through commercial trade channels and in different size classifications, stress grades, and appearance grades.

Lumber includes some strong pieces, many pieces of adequate strength, and some weak pieces. Visually inspect and sort lumber to select the stronger pieces. Discard pieces that show rot or are noticeably undersized. The stronger pieces can be used for joists and rafters and the weaker pieces for plates, sills, braces,

Table 3-1. Nominal and minimum dressed lumber sizes.
Lumber is available in nominal and dressed lumber sizes. Thicknesses apply to all widths; widths apply to all thicknesses. Dressed sizes are for dry lumber.

Item	Thickness Nominal	Thickness Dressed	Face widths Nominal	Face widths Dressed
	- - - - - - - - - - - - - in. - - - - - - - - - - - -			
Boards	1	¾	2	1½
	1¼	1	4	3½
	1½	1¼	6	5½
			8	7¼
			10	9¼
Dimension	2	1½	2	1½
	3	2½	4	3½
	4	3½	6	5½
			8	7¼
			10	9¼
			12	11¼
Timbers (Dressed green)	5 and thicker	½ off	5 and wider	½ off

subflooring, pens and partitions, and other nonstructural uses.

Commercial lumber is classified by size or use categories. Grading rules that determine allowable stresses for design use member size as a guide. For example, a rectangular cross section is more efficient as a beam than a nearly square cross section member. Available size classifications are:

- Dimension lumber (2"-5" thick, 2" or more wide). Including light framing (2"-4" thick, 2"-4" wide), joists and planks (2"-4" thick, 6" and wider), studs (2"-4" thick, 2"-4" wide), and appearance (2"-4" thick, 2" and wider).
- Timbers (5" or more in least dimension).
- Boards (up to 1½" thick, 2" or more wide)

The allowable stresses of lumber are indicated by stress grades. The user must understand the relationship between the size category (e.g. 1" thick, 2" thick, etc. and/or 4" wide, 6" wide, etc.) and stress grade (e.g. No. 1, No. 2, Construction, etc.) because **different** allowable stresses apply to the **same** stress grades in **different** size categories. For example, a Construction grade 2x4 has an allowable bending stress of 1,000 psi and a No. 2 grade 2x4 has an allowable bending stress of 1,400 psi. Refer to the *National Design Specification* (NDS) for allowable stress values. Common stress grades are:

- Select Structural—where high strength and stiffness are needed.
- No. 1—same as Select Structural except lower in quality.
- No. 2—for general construction.
- No. 3—for general construction where appearance and strength are not major factors.
- Construction—for general framing.
- Standard—for lower quality general framing.
- Utility—for blocking, plates, etc.
- Economy—for crates and temporary construction.
- Stud—specialized grade for stud applications.

- Appearance—similar to No. 1, used for high strength and fine appearance in housing and light construction.

WCLB = the grading association.
MILL 10 = the mill code number.
CONST = the lumber grade.
DOUG FIR = the lumber species.
S-DRY = lumber surfaced at 19% mc or less.

S-GRN = lumber surfaced green.
DOUG FIR-L = the species is mixed Douglas fir and larch.
WWP = the grading association.
12 = the mill code number.
SEL STR = the lumber grade.

Fig 3-3. Typical grade stamps.

Wood preservatives

Creosote and pentachlorphenal are oil borne preservatives. ACA preservative treated lumber and CCA are common water borne wood preservatives. Federal regulations are beginning to limit use of preservatives. Generally, keep creosote and pentachlorphenal outdoors or cover it; do not use where feed or water is stored.

Use pressure preservative treated lumber if it will be exposed to soil, insects, manure, weather, or humid conditions. Brushing or dipping is not adequate. Select treatment according to the expected exposure, Table 3-2.

Insist on a grade stamp on each piece of lumber showing the treatment plant and treatment level. It is not possible to determine treatment level from visual inspection.

Treat cut pressure preservative treated wood with a double-brush treatment of copper naphthanate. Salt preservatives tend to accelerate rusting and corrosion of ordinary steel nails. Use double hot dipped galvanized or stainless steel nails or other noncorrosive fasteners.

Wear protective gloves and clothing when handling treated wood. When sawing, cutting, sanding, planing, or boring, use goggles and dust masks. Dispose of treated lumber scraps by burying. Do not burn treated lumber.

Plywood and Particle Boards

Design, selection, and use of plywood has become more complex. To be assured of consistent quality, plywood should have the American Plywood Association (APA) trademark and the grademark of a recognized lumber grading agency. Refer to the APA trademark-grade stamp for plywood specification information, Fig 3-4.

Plywood type, interior or exterior, depends on the resistance of the panels to weather. Exterior-type plywood has better resistance to moisture and weather. Some interior-type plywood is marketed with exterior or intermediate glue but is not adequate for exposed conditions.

Use exterior-type plywood outdoors, inside animal shelters, or where wetting and drying may occur. It has waterproof gluelines and all plies are at least grade C. Sheathing has waterproof gluelines and some grade D plies. Do not use sheathing outdoors unless it is covered with roofing or siding.

Particle board and wafer or chip board are not recommended for exposed use unless specifically rated as indicated on the grade stamp.

Glue

A glued joint is usually stronger than the wood it holds together. Use dry, smooth wood free of dirt, oil, and other coatings. Most purchased lumber has been planed and is sufficiently smooth. Clean off dirt, paint, and other coatings from the wood. Do not use wood with oil or grease at joint locations.

Generally, preservative-treated wood must be planed prior to gluing for maximum holding power. Wood treated with oil base preservatives tends to bleed; buy wood that has been steamed or otherwise cleaned until bleeding has stopped.

Two glues are recommended for buildings and livestock equipment: Casein and Resorcinol Resin. Follow manufacturer's recommendations.

- Resorcinol Resin can be used for both wet and dry conditions. Buy a glue that meets Military Specification Mil-A-46051. Apply glue at 70 F or above. Assemble but wait 5 to 10 min before applying pressure. Maintain pressure for 10 to 16 hr.
- Use Casein only for dry or occasionally wet conditions. Buy a glue that meets Federal Specification MMM-A-125 Type II. Apply glue at 40 F or above; 70 F is recommended. Apply pressure as soon as possible. Maintain pressure for two days at 40 F, 4 hr at 70 F, or 2 hr at 80 F.

Pressure squeezes glue into a thin continuous film, forces air from the joint, brings glue and wood together, and holds them until the glue has set and cured. Use clamps, nails, screws, or other fasteners

Table 3-2. Minimum preservative retention levels.
Adapted from *1985 ASAE Standards.*

	Creosote lb/ft³	Penta-chloro-phenol lb/ft³	ACC lb/ft³	ACA or CCA lb/ft³
Poles, round structural members				
Southern pine, Ponderosa pine	7.5	0.38	NR	0.60
Red pine	10.5	0.53	NR	0.60
Coastal Douglas fir	9.0	0.45	NR	0.60
Jack pine, lodgepole pine	12.0	0.60	NR	0.60
Western red cedar, western larch intermountain Douglas fir	16.0	0.80	NR	0.60
Posts, structural member, All softwood species Sawn four sides	12.0	0.60	NR	0.60
Posts, fence All softwood species Round, half-round, and quarter-round	0.80	0.40	0.50	0.40
Sawn four sides	10.0	0.50	0.62	0.50
Lumber All softwood species In contact with soil	10.0	0.50	0.62	0.50
Not in contact with soil	8.0	0.40	0.25	0.25

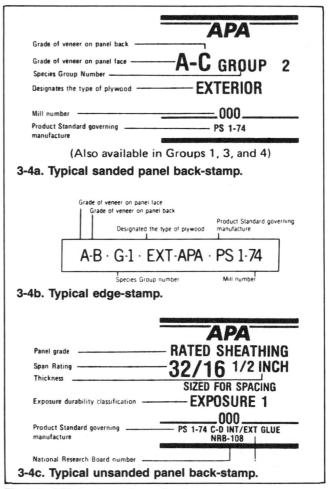

3-4a. Typical sanded panel back-stamp.

3-4b. Typical edge-stamp.

3-4c. Typical unsanded panel back-stamp.

Fig 3-4. APA trademark-grade stamp.

Nails

Table 3-3. Size and strength of common nails.

Size	Length in.	Dia. in.	Approx. no./lb	Approx. strength lb	
				Pull[a]	Lateral[b]
Common nails					
2d	1	0.072	847	Douglas fir,	
3d	1¼	0.080	543	larch or	
4d	1½	0.099	294	southern pine	
5d	1¾	0.099	254		
6d	2	0.113	167	29	63
7d	2¼	0.113	150		
8d	2½	0.131	101	34	78
9d	2¾	0.131	92		
10d	3	0.148	69	38	94
12d	3¼	0.148	63	38	94
16d	3½	0.162	49	42	107
20d	4	0.192	31	49	139
30d	4½	0.207	24	53	154
40d	5	0.225	18	58	176
50d	5½	0.244	14	63	202
60d	6	0.263	11	68	223
Spikes					
10d	3	0.192	32	49	139
12d	3¼	0.192	31	49	139
16d	3½	0.207	24	53	155
20d	4	0.225	19	58	176
30d	4½	0.244	14	63	202
40d	5	0.263	12	68	223
50d	5½	0.283	10	73	248
60d	6	0.283	9	73	248
⁵⁄₁₆	7	0.312	6	80	289
³⁄₈	8-12	0.375	5-3	96	380
Hardened threaded nails					
6d	2	0.120	190	80	69
8d	2½	0.120	117	90	82
10d	3	0.135	78	100	94
12d	3¼	0.135	73	100	94
16d	3½	0.148	57	110	107
20d	4	0.177	36	135	139
30d	4½	0.177	31	135	139
40d	5	0.177	27	135	139
50d	5½	0.177	23	135	139
60d	6	0.177	18	135	139

[a]Per inch penetration of point.
[b]For penetration of 11 diameters.

Table 3-4. Nail selection.

Use galvanized nails where corrosion and staining can occur. Use annular and spiral grooved nails for structural joints. Cement coated nails increase joint strength for a short time, but strength drops to that of a plain nail joint in a few months. Use aluminum nails with aluminum sheets.

The nailing strength of c-c or better plywood is about the same as solid wood, but the greater resistance to splitting when nailed near edges is a definite advantage.

"d" means penny.

	Nail to use
1" thick lumber	8d
2" thick lumber	16d to 20d
3" thick lumber	40d to 60d
Concrete forms	Common or double headed nails.
Toenailing 2x4s, studs, etc.	10d
Sheathing roof, wall, and floor	8d
Roofing	
Aluminum	1¾"-2½" aluminum nail with rubber washers.
Asphalt shingles	Large head roofing nail
Wood shingles	3d to 4d
Steel sheet metal (roofing and siding)	Self-tapping screws, helical drive screws (lead washers for galvanized metal; neoprene for painted)
Fastening to concrete	Concrete nails, helical nails or impact anchor screws.
Plywood—structural work	
Combined subfloor and underlayment: ¾" or less	6d deformed shank
⅞" or larger	8d deformed shank
Subflooring: ⅞" or less	8d common
1" or larger	10d common
Underlayment	3d ring shank
Sheathing: ½" or less	6d common, ring shank
½" or less	or spiral thread
⅝" to 1"	8d common, ring shank or spiral thread

before wiping off excess glue. Use box, galvanized, or cement coated nails—one per 8 in² of joint. Do not remove the nails after the glue has cured.

Concrete

Concrete is a mixture of Portland cement, water, and aggregates. Cement and water form a paste that hardens and glues the aggregates together. Concrete must be proportioned for the intended use and properly mixed, placed, finished, and cured.

Concrete is usually sold by the cubic yard (27 ft³) and price is based on the compressive strength of the concrete. Compressive strength ranges from 2,000 psi to over 4,000 psi. Higher strength concrete has a higher proportion of cement to aggregate and costs more.

Concrete Strength and Durability

Concrete strength and durability depend on the quality of materials used, water-cement ratio, placing, finishing, and curing. To maintain concrete quality and durability, control the amount of water in the mix. Add just enough water to maintain workability. Additional water reduces the strength and durability of the concrete. Concrete with a low water-cement ratio will be more watertight for greater durability both to acid and traffic. Admixtures or additives can increase or decrease set time, increase strength, improve workability, or entrain air.

Power plant fly ash is being added to concrete mixes to decrease costs and in some cases to improve

Table 3-5. Concrete mixes.
Make a trial batch to check for slump and workability. Wet sand occupies more volume than dry sand, so yield is not quite the same as in the example below.

| | Max. size aggregate | [a]Gallons of water for each 94 lb sack of cement, using: | | | [b]Suggested mixture for 1-sack trial batches | | | 94 lb sacks cement per cubic yard[f] |
		Damp[c] sand	Wet[d] (average) sand	Very[e] wet sand	Cement, 94 lb sacks ft³	Aggregates Fine ft³	Coarse ft³	
5-gallon mix (>4,000 psi): use for concrete subjected to severe wear, weather, or weak acid and alkali solutions.	¾"	4½	4	3½	1	2	2¼	7¾
6-gallon mix (3,500-4,000 psi): use for floors (home, barn), driveways, walks, septic tanks, storage tanks, structural concrete.	1"	5½	5	4½	1	2¼	3	6¼
	1½"	5½	5	4½	1	2½	3½	6
7-gallon mix (2,500-3,000 psi): use for foundation walls, footings, mass concrete, etc.	1½"	6¼	5½	4¾	1	3	4	5

[a]Increasing the proportion of water to cement reduces the strength and durability of concrete. Adjust the proportions of trial batches without changing the water-cement ratio. Reduce gravel to improve smoothness; reduce both sand and gravel to reduce stiffness.
[b]Proportions vary slightly depending on gradation of aggregates.
[c]Damp sand falls apart after being squeezed in the palm of the hand.
[d]Wet sand balls in the hand when squeezed, but leaves no moisture on the palm.
[e]Very wet sand has been recently rained on or pumped.
[f]Medium consistency (3" slump). Order air-entrained concrete for outdoor use.

strength. Using fly ash in concrete changes its characteristics and requires careful mixing, installation, and curing because of the variability of fly ash from different sources. Fly ash concrete has been most successful when used in warm weather and with experienced concrete workers.

Required concrete strength is based on the intended use. Concrete subject to freezing and heavy vehicle traffic requires higher strength than building footings.

Concrete in contact with manure and silage or subjected to freeze-thaw cycles must be proportioned for durability. Concrete durability requirements are based on the water-cement ratio unless the strength requirements are greater. Consider strength requirements before ordering concrete, Table 3-6. Concrete floor thickness is also based on use. The heavier and more frequent the loads, the thicker the floor.

Recommended floor thickness:
- 4": feeding floors with minimum vehicle traffic, building floors.
- 5": paved feedlots, building driveways.
- 6": heavy traffic drives (grain trucks, large tractors, and wagons).

For home mixing, see Table 3-5. Consider using a cement that contains an air entraining admixture. Do not use cement with hard lumps that do not break apart—it may be partly set and not make good concrete. Mix concrete for at least one min after all ingredients have been added.

Air-Entrained Concrete

Use air-entrained concrete for all agricultural construction subject to freezing and thawing, manure, or chemicals. Air-entrained

Table 3-6. Durability and strength for air-entrained concrete.
Durability (weathering and chemical resistance) and strength depend primarily on water-cement ratio.
Adding water after the truck arrives rapidly lowers durability and strength. Only ½ gal/bag (about 3 gal/yd) separates the groups in the table.
These are approximate guidelines—follow plans or specifications if available.

Kind of job	Approx. strength psi	Gallons water/bag cement	Water-cement ratio, lb water per lb cement
Feedbunks, slats, above-ground bunker silos	4,500	5.0	0.44
Unventilated manure tanks, parking lots, underground silos	4,000	5.5	0.49
Feedlots, floors, walls, drives, basements; open top or ventilated manure tanks, reinforced retaining walls, beams and columns	3,500	6.0	0.53
Footings, foundation walls, gravity retaining walls	3,000	6.5	0.62

Table 3-7. Aggregate size and air entrainment.
With aggregates well graded up to a larger maximum size, less cement is needed, less air entraining is needed, and the mix is more economical.

Maximum aggregate size, in.	Average air entrainment, %
1½	5
1	6
¾	6
½	7
⅜	7½

concrete has an ingredient added to purposely entrain microscopic air bubbles in the concrete. Entrained air bubbles dramatically improve the durability of concrete exposed to moisture and freezing and thawing cycles. As the water in concrete freezes, it expands, causing pressure that can rupture concrete. Entrained air bubbles act as a reservoir, relieving the pressure and preventing damage to the concrete. Air entrainment also improves the workability of fresh concrete and reduces the amount of bleed water that rises to the surface.

The largest practical size of coarse aggregate is:
- About ⅕ the thickness of vertically formed concrete.
- About ⅓ the thickness of flat concrete work such as floors or walks.
- For reinforced concrete, not over ¾ the distance between bars, or between bars and forms. For example, if reinforcing steel is 1″ from the face of the slab, use ¾″ maximum aggregate.

Concrete Floor Construction

Remove all sod and organic matter from the site. The subgrade must have uniform soil compaction and moisture content and be well drained. The top 6″ of subgrade should be sand, gravel, or crushed stone to provide for drainage under the slab. This is especially important where the slab will be wet and subject to freezing.

After the concrete has started to stiffen, round the edges to prevent chipping. Shrinkage cracks are unavoidable, so cut control joints ¼ of the slab thickness deep to prevent random cracks. Reducing water in the mix will help reduce shrinkage and cracking. Loads are transferred across the joints by the aggregates in the broken concrete surfaces below the cut. Divide the slab into rectangles:
- 8′x12′ for 4″ thickness.
- 10′x15′ for 5″ thickness
- 12′x18′ for 6″ thickness

Cut control joints in fresh concrete with a pointed trowel or straight hoe, or saw them after the concrete has cured enough for smooth cuts but before the random cracks form.

Isolation and expansion joints permit the slab to move with earth and temperature changes. Place ½″ wide isolation joints along existing improvements such as buildings, concrete water tanks, or paved drives. Install expansion joints in long walks and drives to prevent buckling of the slab during hot weather. Increase the slab edge thickness to help reduce cracking and rodent problems, resist equipment damage, and to allow for soil erosion.

Exercise caution when placing flat slabs or slabs-on-grade. Slabs without a footing and foundation wall can harbor rodents.

Additional information on the mixing and placing of concrete is available in the Midwest Plan Service's MWPS-35, *Farm and Home Concrete Handbook*, and Portland Cement Association's publications *Design and Control of Concrete Mixtures, Cement Mason's Guide*, and *Concrete Paved Feedlots*.

Curing

Cure fresh concrete by covering with a plastic film, continuously ponding with water, covering with wet burlap or straw, or applying a curing compound. Excessive evaporation from the surface of fresh concrete reduces its ultimate strength. Be sure to cure concrete walls by one of these methods after the forms are removed. Proper curing can increase concrete strength about 50% over concrete allowed to air dry after finishing. Begin curing as soon as the concrete surface is hard enough not to be damaged by the water. Continue this for 5 to 7 days. Do not let new concrete freeze for at least 5 to 7 days after placing.

Slip Resistant Concrete Floors

In cattle housing, use a rough concrete surface for floors, aprons, and around waterers. Avoid a steel troweled concrete finish because it is smooth and slippery, especially when wet. Wood float and broom finished surfaces can become smooth in time due to scraping and constant animal traffic. Select the degree of floor roughness based on the intended use and animal type.

New floors can be scored or grooved as they are placed with a homemade tool, Figs 3-5 and 3-6. Make grooves ⅜″-½″ deep by ½″-1″ wide, spaced 4″-8″ apart. Make grooves diagonal to the direction of animal traffic. Parallel grooves can be used where floors are flushed. Do not make grooves perpendicular to animal traffic because the floor is harder to scrape. A diamond groove pattern can be used around waterers, cattle chutes, and walkways for better traction, Fig 3-7. Deep grooves make cleaning and disinfecting more difficult and can cause foot and leg problems with smaller animals.

Where grooves present a problem, aluminum oxide can be added to the surface when the floor is placed. Apply aluminum oxide grit (as in sandpaper) at ¼ to ½ lb/ft² and tamp into the finished surface before the concrete sets. Coarse grit (4 to 6 mesh/in) is recommended.

Existing slick floors can be grooved with a mechanical grinder (similar to ones used for sawing concrete).

Badly worn but sound floors can be resurfaced with a concrete overlay. Bonding a thin overlay to an existing floor requires special cleaning and concrete mixes. Refer to the Portland Cement Association publication, *Resurfacing Concrete Floors*, for additional information on floor resurfacing.

For more information about slip resistant floors, refer to Midwest Plan Service's MWPS-35, *Farm and Home Concrete Handbook*.

Insulation

Insulation is any material that reduces heat transfer from one area to another. Although all building materials have some insulation value, the term insulation usually refers to materials with a relatively high resistance to heat flow. The resistance of a material to heat flow is indicated by its **R-value.** Good insulators have high R-values. See Table 3-8.

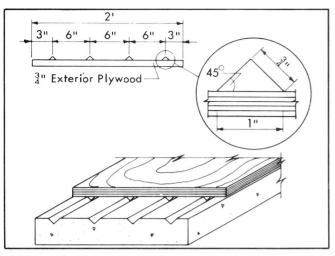

Fig 3-5. Wooden concrete groover.

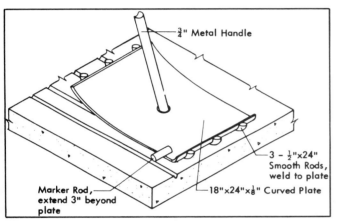

Fig 3-6. Steel concrete groover.

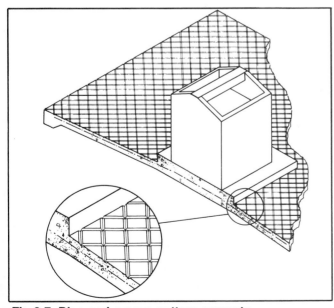

Fig 3-7. Diamond groove pattern concrete.

Table 3-8. Insulation values.

Adapted from *1981 ASHRAE Handbook of Fundamentals*. Values do not include surface conditions unless noted otherwise. All values are approximate.

Material	R-value Per inch (approximate)	R-value For thickness listed
Batt and blanket insulation		
Glass or mineral wool, fiberglass	3.00-3.80	
Fill-type insulation		
Cellulose	3.13-3.70	
Glass or mineral wool	2.50-3.00	
Vermiculite	2.20	
Shavings or sawdust	2.22	
Hay or straw	1.50	
Rigid insulation		
Expanded polystyrene		
Extruded, plain	5.00	
Molded beads, 1 pcf	5.00	
Molded beads, over 1 pcf	4.00-4.30	
Expanded rubber	4.55	
Expanded polyurethane, aged	6.25	
Glass fiber	4.00	
Wood or cane fiberboard	2.50	
Polyisocyanurate	7.04*	
Foamed-in-place insulation		
Polyurethane	6.00	
Urea formaldehyde	4.00	
(not recommended)		
Building materials		
Concrete, solid	0.08	
Concrete block, 3 hole, 8″		1.11
Lightweight aggregate, 8″		2.00
Lightweight, cores insulated		5.03
Metal siding	0.00	
Hollow-backed		0.61
Insulated-backed, ⅜″		1.82
Lumber, fir and pine	1.25	
Plywood, ⅜″		0.47
Plywood, ½″		0.62
Particleboard, medium density	1.06	
Hardboard, tempered, ¼″		0.25
Insulating sheathing, ²⁵⁄₃₂″		2.06
Gypsum or plasterboard, ½″		0.45
Wood siding, lapped, ½″x8″		0.81
Windows (includes surface conditions)		
Single glazed		0.91
With storm windows		2.00
Insulating glass, ¼″ air space		
Double pane		1.69
Triple pane		2.56
Doors (exterior, includes surface conditions)		
Wood, solid core, 1¾″		3.03
Metal, urethane core, 1¾″		2.50
Metal, polystyrene core, 1¾″		2.13
Floor perimeter (per ft of exterior wall length)		
Concrete, no perimeter insulation		1.23
With 2″x24″ perimeter insulation		2.22
Air space (¾″-4″)		0.90
Surface conditions		
Inside surface		0.68
Outside surface		0.17

*Time aged value for board stock with gas barrier quality aluminum foil facers on two major surfaces. A ¾″ air space on each side is required.

For more information on insulation types and selection, obtain Midwest Plan Services's AED-13, *Insulation and Heat Loss.*

Insulation Levels

The insulating value of a wall or ceiling is the total of the insulation, siding, lining, surface conditions, and air spaces. During the winter, insulation reduces heat transfer and helps control condensation and frost on indoor surfaces. It also helps control summer heat loads.

Minimum insulation levels for warm barns are based on winter heat loss. Warm barns are kept at 40 F or above by good construction, careful control of the ventilating system, and supplemental heat as needed. Provide a minimum insulation level of R=11 in the walls and R=19 in the ceiling for warm buildings such as an office or treatment area. See Table 3-9 and Fig 3-8 for suggested insulation levels based on heating degree days.

Installing Insulation

Figs 3-9 and 3-10 show common construction methods for insulated roofs and walls. Cracks around window and door frames, pipes, and wires result in cold spots, drafty areas, or condensation, and reduce ventilating inlet effectiveness in mechanically ventilated buildings. Caulk cracks and joints on outside surfaces.

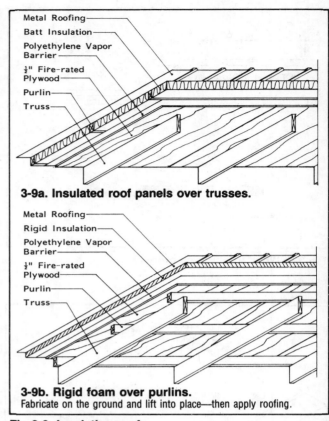

3-9a. Insulated roof panels over trusses.

3-9b. Rigid foam over purlins.
Fabricate on the ground and lift into place—then apply roofing.

Fig 3-9. Insulating roofs.
Batt or rigid type insulation can be used. Consider installation labor and material costs.

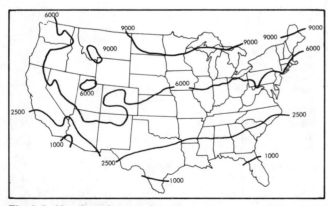

Fig 3-8. Heating degree days.
Accumulated difference between 65 F and average daily temperature for all days in the heating season.

Table 3-9. Minimum warm barn insulation levels.
R-values are for building sections. In cold barns with mature animals, no insulation is needed in the walls or ceiling, but a total R-value of 3 or more in the ceiling controls roof condensation and frosting and reduces summer heat load.

Heating degree days	Minimum R-values Warm barn		
	Walls	Ceiling	Perimeter
2,500 or less	11	22	4
2,501-6,000	11	25	6
6,001 or more	16	33	8

Maximizing Insulation Effectiveness

Vapor barriers

Wet insulation increases heat loss and building deterioration. Vapor barriers are important in restricting moisture migration through walls, ceilings, and roofs. Vapor barriers are classified by a permeability rating measured in perms. A good vapor barrier has a perm rating less than 1.

Install 4 to 6 mil polyethylene vapor barriers on the warm side of all insulated walls, ceilings, and roofs. With rigid insulation, use tongue and grooved panels or seal joints with caulk or tape to prevent moisture migration. Use polyethylene vapor barriers underneath concrete floors and foundations to control soil moisture. Use waterproof rigid insulation if in contact with soil. Refer to Figs 3-9 and 3-10 for proper vapor barrier locations.

Fire protection

The rate at which fire moves through a room depends on the interior lining material. Many plastic foam insulations have high flame spread rates. To reduce risk with these materials, protect them with fire-resistant coatings. Materials that provide satisfactory protection include:
- Fire rated gypsum board. Use water-resistant type in high moisture environments such as animal housing.

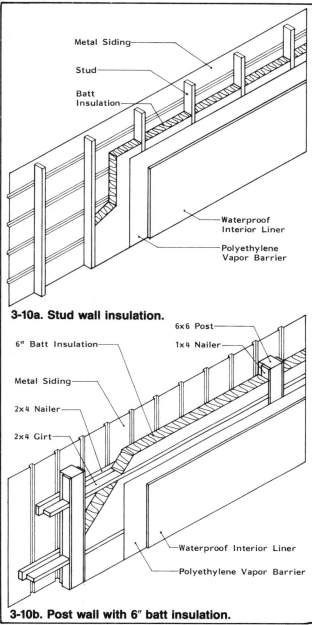

3-10a. Stud wall insulation.

3-10b. Post wall with 6″ batt insulation.

Fig 3-10. Insulating walls.

Table 3-10. Permeability of building materials.
Adapted from the *1981 ASHRAE Handbook of Fundamentals.* A vapor barrier should have a perm rating of less than 1.0.

Material	Perms (dry cup)
Aluminum foil, 1-mil	0.0
Polyethylene plastic film, 6-mil	0.06
Kraft and asphalt laminated building paper	0.3
Two coats of aluminum paint (in varnish) on wood	0.3-0.5
Three coats exterior lead-oil base on wood	0.3-1.0
Three coats latex	5.5-11.0
Expanded polyurethane, 1″	0.4-1.6
Extruded expanded polystyrene, 1″	0.6
Tar felt building paper, 15 lb	4.0
Structural insulating board, uncoated, ½″	50.0-90.0
Exterior plywood, ¼″	0.7
Interior plywood, ¼″	1.9
Tempered hardboard, ⅛″	5.0
Brick masonry, 4″	0.8
Cast-in-place concrete wall, 4″	0.8
Glazed tile masonry, 4″	0.12
Concrete block, 8″	2.4
Metal roofing	0.0

- ½″ thick cement plaster.
- ¼″ thick sprayed-on magnesium oxychloride (60 lb/ft³).
- Fire rated ½″ exterior plywood.

Birds and rodents

To prevent bird and rodent damage, cover exposed insulation with a protective lining and construct buildings so birds cannot roost near the insulation. An aluminum foil covering is not sufficient protection because of joints, cut surfaces, and tears. In warmer climates, consider screening all vent openings. Use ½″x½″ hardware cloth for air intake vent openings and ¾″x¾″ hardware cloth for air outlet vent openings. In cold climates, screened vent outlets are impractical because they freeze shut. Routinely remove and destroy bird nests to reduce bird populations.

4. VENTILATION

Ventilation is an air exchange process that:
- Brings outside air into a building.
- Moves air through all areas of the building to supply oxygen and pick up odors, heat, moisture, dust, and pathogenic organisms.
- Removes contaminated air from the building.

The purpose of ventilation is to provide a healthful environment. Base ventilation design on animal requirements. If operator comfort is considered, adjustments in the ventilating system can be made, but the resulting environment may not be best for livestock.

Ventilating systems can be either natural or mechanical. Natural ventilation utilizes wind pressure and differences between inside and outside temperature to move air through the building. Mechanical ventilation uses fans, controls, and air inlets and outlets to provide a positive air exchange. Natural ventilation is most common in beef buildings. Mechanically ventilate treatment areas, offices, and other warm, insulated rooms. Assist a natural ventilating system that does not adequately ventilate during certain weather conditions with mechanical ventilation.

Natural Ventilation

Wind pressure and the difference between inside and outside air density due to temperature moves air through the building. Natural ventilation of gable buildings works best with a continuous ridge opening, large sidewall openings, no ceiling, smooth roof underside, small continuous eave openings, and a 4:12 roof slope, Fig 4-1.

Wind across the open ridge, along with inside and outside temperature differences, draws warm, moist air out through the ridge and fresh air in through continuous sidewall or eave openings. Locate winter air inlets high on the sidewall or in the eaves to prevent cold drafts on animals. Downwind air inlets will occasionally act as air outlets. Some ventilation occurs even on calm days, because warm air rises, causing a chimney effect.

Large openings (typically ½ or more of the sidewalls) are important to allow a cross flow of air and good hot weather ventilation. Locate openings in the animal zone (first 4' above floor).

Design

Building location

Locate buildings on high ground where trees; upright, tower, or bunker silos; grain bins; and other structures do not disturb airflow. Large, tall, solid obstructions within 100' of buildings can cause ventilation problems during hot weather.

Building orientation

Building orientation affects the performance of a natural ventilating system. Greater wind pressure differences occur when air strikes the side of a building rather than the endwall. This is especially important in warm weather to maximize cross airflow. Because air enters at the eave or sidewall openings and discharges at the ridge, orient buildings so winter prevailing winds are perpendicular to the ridge.

In most of the Midwest, build open-front buildings open to the south with the long axis east-west for the best summer cooling, winter sun penetration, and winter wind control. An open front toward the southeast is suitable to fit existing site conditions and take advantage of prevailing south-southeast summer breezes.

Ridge openings

Provide a ridge opening of 2" (measured horizontally) for each 10' of building width, e.g. a 14" wide opening for a 70' wide building.

To protect against weathering, cover each rafter or truss with flashing. Extend the flashing ½"-1" on each side of the exposed rafter, Fig 4-2.

With a properly sized ridge opening, precipitation entering the building is not a serious problem. Air exiting through the ridge prevents most rain and snow from entering. Ridge caps are not recommended because they disturb airflow, can trap snow, are expensive, and require maintenance.

If the amount of precipitation coming through the ridge is objectionable, it is better to protect areas below the ridge than to build a cap. Cover critical components, such as a feeder motor or feeder belt, or place an internal gutter 2'-3' below the open ridge to collect rain water and channel it out of the building, Fig 4-4.

A 12" ridge vent upstand, Fig 4-3, can help keep out wind and snow and increase the chimney effect when wind is perpendicular to the ridge. Upstands can improve airflow on buildings with less than a 4:12 roof slope.

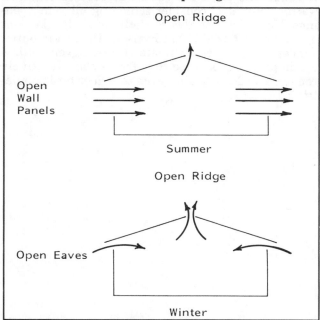

Fig 4-1. Naturally ventilated barn features.

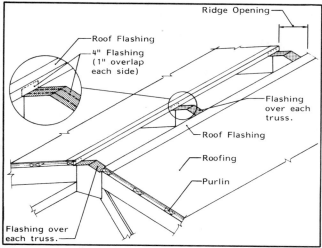

Fig 4-2. Flashing for ridge openings.

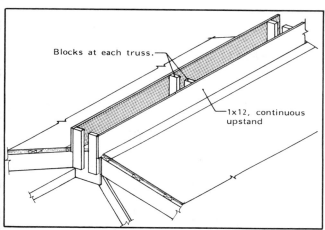

Fig 4-3. Upstand.

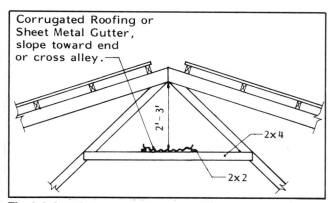

Fig 4-4. Interior trough with an open ridge.

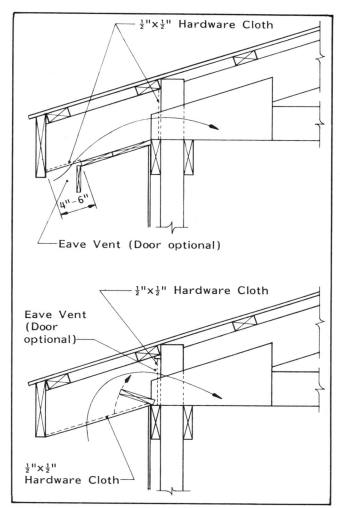

Fig 4-5. Eave openings.
Make the length of eave vent doors about 75% of the opening length.

Eave openings

Construct continuous eave openings along both sides of the building. Size **each** opening to have at least ½ as much open area as the ridge opening. Eave openings can be between trusses or under the roof overhang, Fig 4-5.

In windy, exposed locations, provide eave baffles so airflow can be reduced, but never completely close the baffles. To reduce drafts, protect eave openings from direct wind gusts with fascia boards and vent doors. Use vent doors only during periods of severe winter weather. If vent doors are used, continual adjustment is needed, requiring more labor.

Sidewall openings

Larger sidewall openings are needed for cross ventilation during warm weather. Provide openings in the closed side of open-front buildings and in both long walls in totally enclosed buildings.

Summer vent door types include pivot doors, top- or bottom-hinged doors, and plastic or nylon curtains, Fig 4-6. Provide at least a 4' high continuous opening the length of the building. For good airflow at the animal level, put the lower edge of the opening not more than 4' above the floor. Put vent doors lower if construction details and animal access allow. Build vent doors to open fully so none of the opening is constricted, or increase door size to provide the recommended opening.

Make sidewalls at least 12' high to take advantage of winds for ventilation. See Table 4-1.

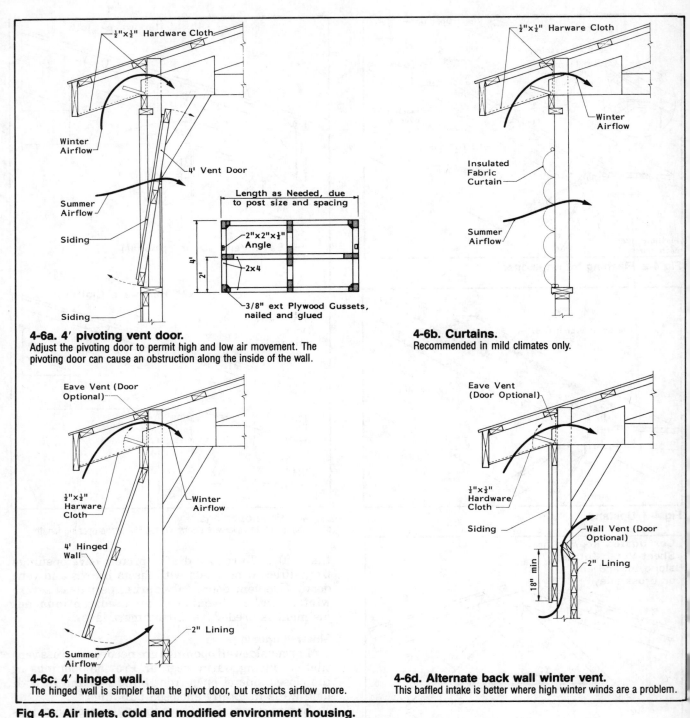

4-6a. 4′ pivoting vent door.
Adjust the pivoting door to permit high and low air movement. The pivoting door can cause an obstruction along the inside of the wall.

4-6b. Curtains.
Recommended in mild climates only.

4-6c. 4′ hinged wall.
The hinged wall is simpler than the pivot door, but restricts airflow more.

4-6d. Alternate back wall winter vent.
This baffled intake is better where high winter winds are a problem.

Fig 4-6. Air inlets, cold and modified environment housing.
Eave vent doors are optional. Do not completely close eave vent doors because of moisture and odor buildup.

Table 4-1. Natural ventilating openings.
Ridge openings are sized for a 10 mph wind.
Sidewall openings are sized for a 2 mph wind and are a minimum. Opening more of the back sidewall increases summer cooling.

Opening	Building width (ft)					
	30	40	50	60	70	80
Ridge opening (in)	6	8	10	12	14	16
Eave openings (in)	3	4	5	6	7	8
Sidewall opening (in)	28	36	44	52	60	68
Sidewall height (ft)	12	12	14	14	16	16

Endwalls
Provide ventilating openings in the endwalls of wide buildings. Endwall doors may be adequate. Provide openable ventilating panels or wall sections in buildings with no large end doors.

Roof slope
Because warm air rises, steeper sloped roofs provide better upward warm airflow. However, with roof slopes over 6:12, incoming air rises rapidly along the

roof and does not drop into the animal zone. Condensation and high interior summer temperatures are a problem with roof slopes less than 4:12 because of reduced air movement. Therefore, for livestock housing use a 4:12 to 6:12 roof slope. An exception is a monoslope building with an 18'-24' high front wall. A 2½:12 roof slope appears to be adequate based on limited field observations.

Management
Winter
The temperature within a naturally ventilated building closely follows the outdoor temperature. Use heated or frost protected water bowls and pipes.

Poor winter ventilation causes animal health problems. Fog, condensation, or frost form when the building is not properly ventilated. Attempts to warm the building by closing eave inlets, wall openings, or ridge openings with baffles increases the problem.

Summer
Hot, muggy summer weather is one of the most critical times for animal health, comfort, and productivity. Often there is little air movement and sunshine on the roof radiates more heat to the animals.

Provide recommended sidewall openings for good cross ventilation to reduce animal heat stress. Adequate ridge and eave openings are important for proper air movement through the building. Low, wide buildings and buildings with flat ceilings or low pitched roofs are difficult to naturally ventilate because adequate openings cannot be provided.

Modify existing buildings with poor ventilation by removing an existing ceiling, opening a ridge, or cutting large sidewall openings for better natural air movement. If summer heat stress is still a problem, use circulation fans to move air over the animals.

Mechanical Ventilation

A mechanical ventilating system has fans, controls, fresh air inlets, and outlets. A well designed system provides greater control over room temperature and air movement than natural ventilation. Mechanical ventilation is used in treatment areas, offices, and confined feeding barns that are kept above freezing.

Ventilating Rates

An effective ventilating system distributes and mixes the air within the building to control temperature, moisture, odor, and pathogenic organisms. The system can be an exhaust (negative pressure) or positive pressure type; however, recommended ventilating rates are the same, Table 4-2. Table values are airflow quantities through the building. Positive pressure systems require duct work and careful installation. Consult your ventilation equipment dealer for more information. Use recirculation systems cautiously because they do not improve air quality.

Table 4-2. Beef ventilating rates.
Size the system based on total building capacity. Table values are additive, e.g. for cows, mild weather requires 50 + 120 = 170 cfm/cow. An alternative hot weather ventilating rate is equal to the building volume (minimum of 1 air change/minute).

	Ventilating rates		
	Cold weather	Mild weather	Hot weather
	- - - - - cfm/animal - - - - -		
Calves, 0-2 mo	15	+35=50	+50=100
Feeder calves, 2-12 mo	20	+40=60	+70=130
Yearlings, 12-24 mo	30	+50=80	+100=180
Cow, 1,400 lb	50	+120=170	+300=470

Example 4-1:

Calculate the cold, mild, and hot weather ventilating rates for a 200-head feeder calf barn. Table 4-2 gives the following ventilating rates:

Cold weather: 20 cfm/head × 200 head = 4,000 cfm total

Mild weather: 40 cfm/head × 200 head = 8,000 cfm (Total = 4,000 + 8,000 = 12,000 cfm)

Hot weather: 70 cfm/head × 200 head = 14,000 cfm (Total = 12,000 + 14,000 = 26,000 cfm)

Temperature Control

To maintain constant room temperature, heat **produced** by animals, electric equipment, and heaters has to equal heat **lost** through building surfaces and ventilation. Calving barns may need supplemental heat supplied by heat lamps or hot air type heaters.

Moisture Control

During cold weather, ventilation brings cold air into the building. The air is warmed by energy from animals, electrical equipment, and supplemental heat. As the air temperature rises and the air expands, its relative humidity decreases and it has a higher moisture holding capacity. The moisture holding capacity of air nearly doubles for every 20 F rise in temperature. Fresh, incoming ventilating air picks up moisture and the fan system exhausts it from the building, Fig 4-7.

Design ventilation to maintain room air at 50%-65% relative humidity. Higher humidities increase condensation and health problems; lower hu-

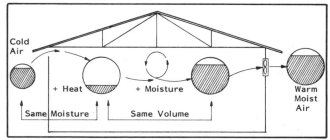

Fig 4-7. Moisture removed by ventilation.

midities increase dust levels. Also, a 50%-65% relative humidity reduces airborne bacteria found in livestock buildings.

Exhaust Ventilation

Exhaust ventilation is most common. Exhaust fans pull air out of the building, creating a partial vacuum or negative pressure inside the building. The pressure difference creates airflow through designed inlets and other openings, Fig 4-8. Size exhaust fans based on Table 4-2.

Air Inlets

Properly located and designed air inlets allow air to flow uniformly to all parts of a building. Usually inlets are designed for the hot weather rate, then reduced with a baffle for lower rates.

Maintain the interior temperature within 10 F of outside temperature to help control condensation. A continuos slot inlet that forces air down the wall into the animal zone is preferred.

For buildings up to 40′ wide, place slot inlets at the ceiling along both sidewalls. For wider buildings, add one or more interior ceiling inlets and put fans in both long walls, Fig 4-11. With interior ceiling slots, insulate the ceiling and roof to control condensation during the winter and reduce excess heating of summer inlet air.

When housing young calves in a room where temperatures are more than 10 F above outside temperature, careful control of inlet air is important. Bring air directly in from the outside year-round to reduce attic moisture problems.

During cold weather, close and seal air inlets within 8′ of the winter exhaust fans to prevent short circuiting.

Fans

Select fans to move enough air against at least ⅛″ static pressure. Variable-speed fans have poor pressure ratings at low speeds. Purchase fans that have an Air Movement and Control Association (AMCA) "Certified Rating" seal or equivalent testing and rating.

Use fans designed specifically for animal housing. Select fans with an epoxy coated fiberglass or stainless steel housing. Painted housings have a short life. Buy totally enclosed, split phase, or capacitor-type farm duty fan motors. Wire each fan to a separate circuit to avoid shutdown of the entire ventilating system if one motor blows a fuse. Protect each fan with a time delay fused switch at the fan. Size time delay fuses at 25%-50% over fan amperage.

Select fan motors with thermal overload protection and manual reset switches. Manual reset switches reduce the safety hazard of a fan starting while being checked and reduce on-off cycling that can damage the motor.

With a properly sized, continuous rigid baffle slot inlet, fan location has little effect on air distribution. Place winter fans in the downwind wall for buildings less than 40′ wide and in both walls for wider buildings. To maintain uniform air inflow to all parts of

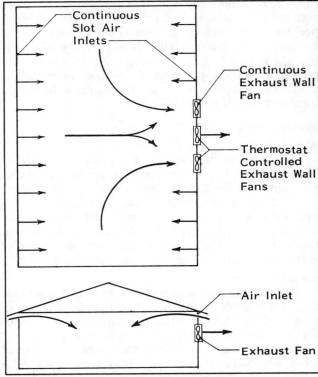

Fig 4-8. Exhaust (negative pressure) ventilation.

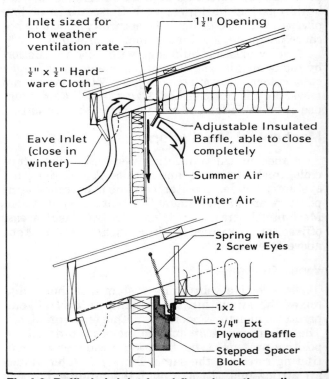

Fig 4-9. Baffled air inlet for airflow down the wall.

the building, place fans no more than 75′ from the farthest inlet. Space mechanically ventilated buildings at least 35′ apart, so fans do not blow foul air into the intakes of adjacent buildings (75′ apart for access and fire protection).

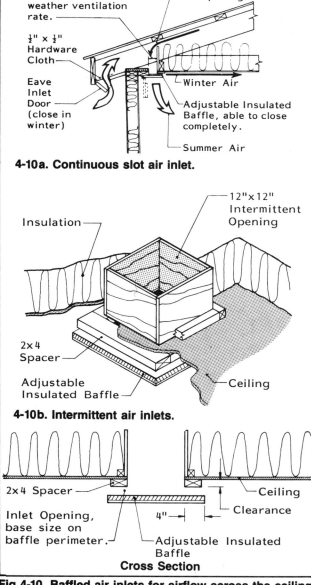

4-10a. Continuous slot air inlet.

4-10b. Intermittent air inlets.

Fig 4-10. Baffled air inlets for airflow across the ceiling.

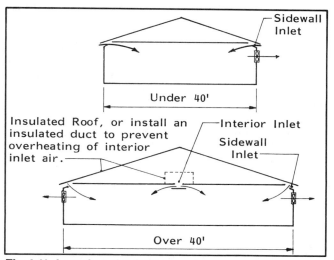

Fig 4-11. Locating slot inlets.

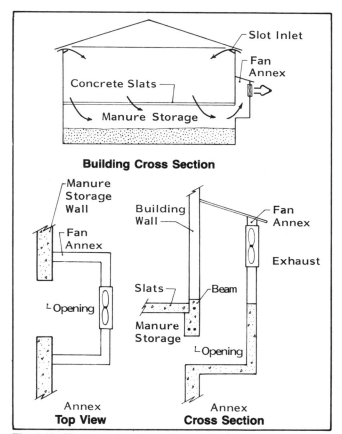

Fig 4-12. Manure storage exhaust fan annex.
Size ventilation openings to supply the cold or mild weather rate. Provide an air velocity of 800 ft/min through the opening.

Manure Storage Ventilation

Ventilation reduces manure gas accumulation in the animal area, reduces odor levels, improves air distribution, and helps warm and dry floors. Provide continuous ventilation from all enclosed in-building and under-building manure storages.

Allow at least 12″ clearance between the bottom of slat support beams and the manure surface. Variable-speed fans are not recommended. Fans with a corrosion-resistant finish are required.

Annexes

A fan annex or "dog house" is common in buildings with a manure storage under the animal space, Fig 4-12. It does not provide air distribution through the floor as uniformly as a duct, but it is less expensive and performs satisfactorily.

Design annex ventilation to supply at least the cold weather rate but no more than the mild weather rate. Locate annexes so no point in the storage is farther than 75′ from an annex.

5. COW-CALF FACILITIES

Traditionally, beef cattle have been produced on rangeland with open grazing. Interest in managing cows and calves on drylot is increasing because producers want to spread out labor; utilize crop residues; provide better year-round conditions; reduce fencing costs, pollution problems, land requirements; and reduce losses due to predators, disease, and weather.

Cows bred to calve in the spring are usually wintered outdoors. Windbreak fences, shelterbelts, haystacks, and natural terrain provide satisfactory shelter. Confine cows near treating facilities before calving for observation. During cold, windy, or wet weather, provide a calving barn for protection during the first few days after birth and for sick cow care. An open-front building for cow shelter, calf creeps, and calf dryer-warmer boxes with hot air blowers aid cold, wet weather calving. Use warmer boxes cautiously, because overheating calves leads to chilling and sickness. Return cows and calves to pasture after the first few days and provide portable calf shelters.

Cow-Calf Feedyards

To feed different rations and improve observation, provide separate feedyard areas for mature cows, first calf heifers, bulls, and calves. Layout variations depend on size, roughage storage, feeding equipment type, drainage, and management. Portable silage and grain bunks are useful for up to 200 head. Use fenceline feedbunks for herds above 100 head and mechanical feedbunks for up to 500 head fed daily in a dry lot.

Figs 5-1 to 5-3 show wintering lot layouts for different herd sizes. Lot size assumes animals winter in the lots and go to pasture in the spring. A cow-calf

unit in a lot year-round may need over 1,000 ft². Required lot space depends on soil type, lot slope, and precipitation. Consider all these factors when planning cow-calf feedyards. The 80 cow-calf layout allows for gradual expansion. The 150 cow-calf feedyard uses more lots to better manage age groups. The 300 cow-calf feedyard layout provides more management flexibility.

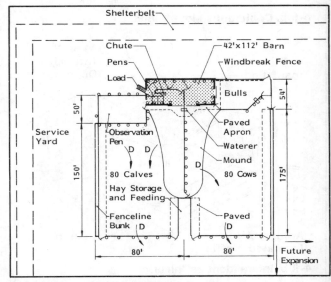

Fig 5-1. 80 cow-calf feedyard layout.
An expandable basic layout for wintering lots. Adjust lot space and dimensions to fit your soil type, seasonal rainfall, and lot slope. Consider prevailing winds when locating protection. One barn is for calving and working facilities. Portable feedbunks can be used initially, and fenceline bunks and drive alleys added later. Add facilities and lots for heifers or more cows during expansion.

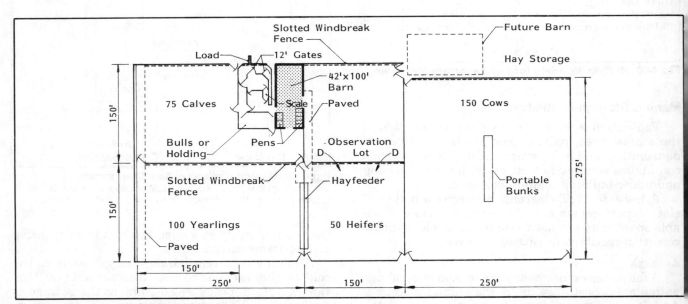

Fig 5-2. 150 cow-calf feedyard layout.
Multiple pens allow smaller groups, less size variation among animals, and more individual feeding. Consider drive alleys and fenceline bunks between the lots for easier feeding and animal handling. Fenceline bunks make it easier to feed during wet weather and to locate and maintain mounds.

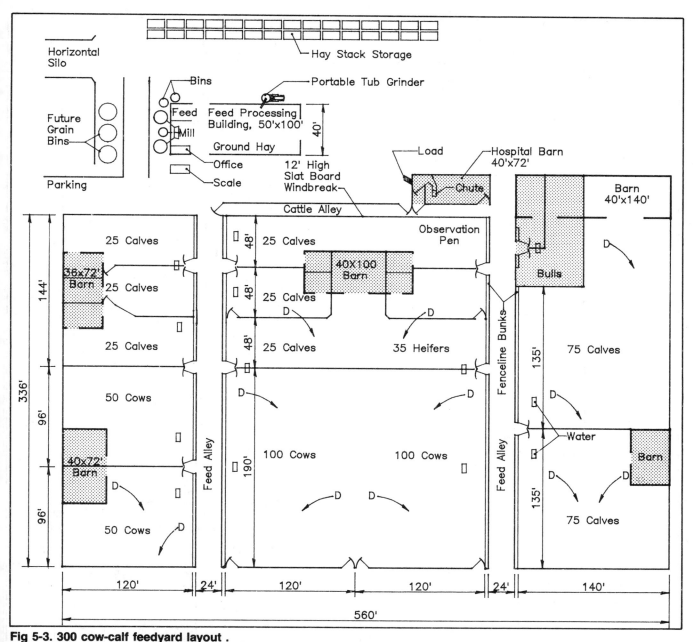

Fig 5-3. 300 cow-calf feedyard layout .
This layout allows for a year-round drylot operation if pavement is provided in pens. Several lots allow for handling groups according to their needs. Locate the feed storage and preparation facility near the feedyard for better control over year-round feeding.

Calving Facilities

Provide clean, well drained, and wind protected pastures for outdoor calving. During cold weather, a calving barn may be needed. To reduce the risk of sickness, a cool, dry, properly ventilated barn is essential. A closed, dark, damp, warm barn can increase disease problems. Provide ventilation to remove moisture and excess heat. See the ventilation chapter.

Calving facilities include a loose housing observation area, individual pens, and a chute for holding and treating cows. Hinged, swing away, or interchangeable panels and gates allow flexibility in indi-vidual and group calf pens and creep arrangements. Move gates for easier cleaning.

A barn that houses 5%-10% of the cow herd is adequate in mild climates. Barn space for 15%-20% of the herd is needed when several cows or heifers require close observation at one time, when there is greater risk of severe weather, or with heat synchronization, artificial insemination, or similar practices. Hold cows close to calving in an observation area and move into mothering pens if necessary.

Mothering pens are usually 10'x10', or sometimes 8'x12'. When possible, 12'x12' pens for larger animals provide more room to assist cows during calv-

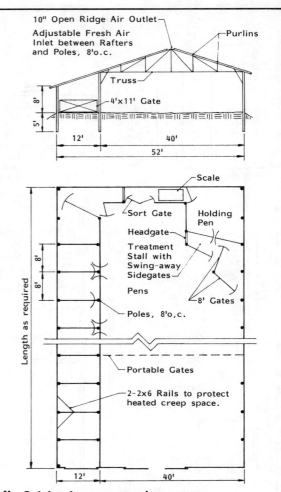

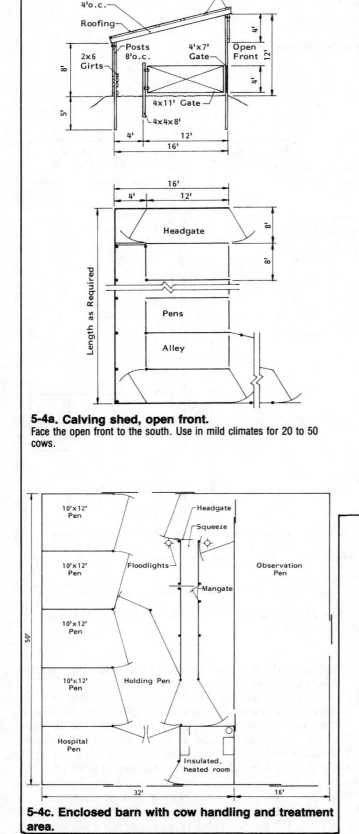

5-4a. Calving shed, open front.
Face the open front to the south. Use in mild climates for 20 to 50 cows.

5-4b. Calving barn, covered area.
Hold cows for observation and handling in a section of the 40' area along the pens. Use in severe climates for herds of several hundred cows.

5-4c. Enclosed barn with cow handling and treatment area.

Fig 5-4. Calving barn arrangements.

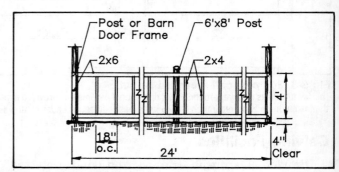

Fig 5-5. Calf creep fence.
Fence across barn doors or pen fronts so calves, not cows, can get to the grain feeders.

ing. Three solid sides on pens reduce drafts and help keep animals separate and quiet. Provide a feed manger for the cow and a lighted, heated creep area for the calf. A frostproof hydrant provides water during winter calving.

Dry newborn calves and warm chilled calves in warmer/dryer boxes kept at 70 to 110 F. See the equipment pages in this chapter for plans and examples. A

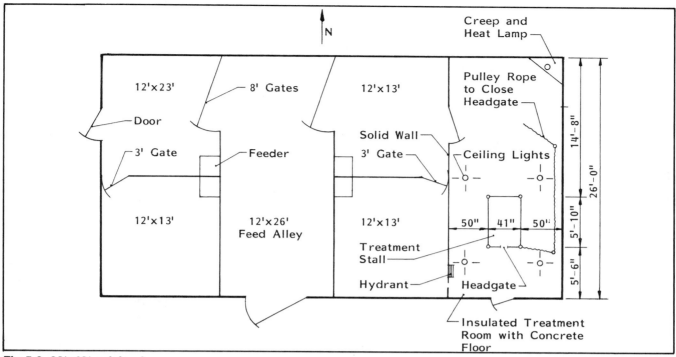

Fig 5-6. 26′x48′ calving barn.
A small herd layout with 4 to 6 calving pens and treatment area.

Pasture Corral Systems

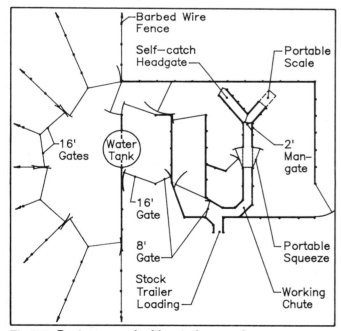

Fig 5-7. Pasture corral with rotation grazing.

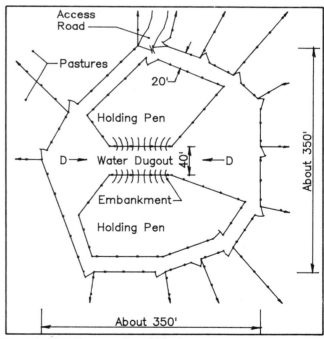

Fig 5-8. Rotation grazing with dugout waterer.

heated hover gradually adjusts calves to winter temperatures. With movable calf shelters, turn cows and calves outdoors in 2 or 3 days.

Artificial Insemination Facilities

Special facilities for artificial insemination (AI) include holding, feeding, and observation pens; a crowding pen; chute; and a breeding or insemination stall. Provide convenient storage for insemination supplies and semen tank.

For artificial insemination use a breeding stall because conception may be affected with usual cattle handling-treatment facilities. A breeding stall is usually a blind alley that restrains the cow from

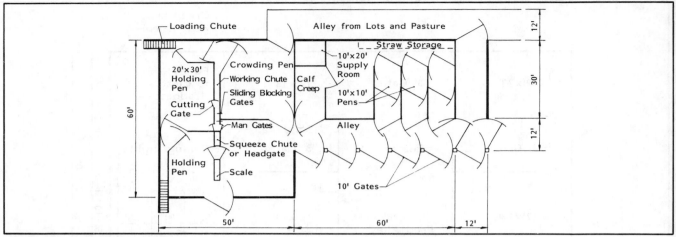

Fig 5-9. Calving barn and corral.
Suitable for several hundred cows.

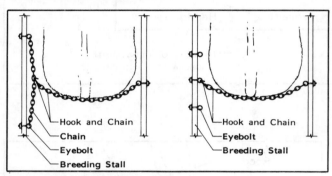

Fig 5-10. Chain fastening methods.

Fig 5-11. Artificial insemination chute.
USDA plan no. 6103.

forward and lateral movement. No single part of the cow is restrained so the cow stands quietly and does not fight to free herself. Benefits of a breeding stall are reduced animal stress, reduced handler frustration, and increased chance of conception.

The front, top, and sides are usually solid to keep cows from being distracted. Use a feed sack or other opaque covering to drape over the rear of the cow to darken the stall. Use a chain or bar across the the rear of the stall high enough to prevent the cow from stepping over and low enough to allow easy access, Fig 5-10. Locate the chain about half way between the hooks and pins of the cow. Fasten a chain so it can be released with one hand. With a bar, place it in slots so it slides forward if the cow moves.

A portable gate-pen-chute permits AI on pasture, which minimizes investment but requires added labor. Weather can be a problem.

A herringbone AI chute allows breeding large herds and is practical for heat synchronization. If you plan to use heat synchronization, consult your veterinarian while planning facilities. Provide for efficiently handling all the cows in a group. Gather cows to be bred in a holding pen and quietly work them through the chute. A chain behind the cow keeps her from backing out. The front, side, and top are solid to keep cows from being distracted. After insemination, the cow exits through a spring-loaded front gate. Adjust stall width for different size animals. With

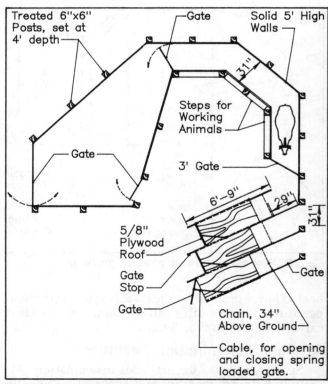

Fig 5-12. Herringbone artificial insemination system.
For servicing larger groups that have been heat synchronized.

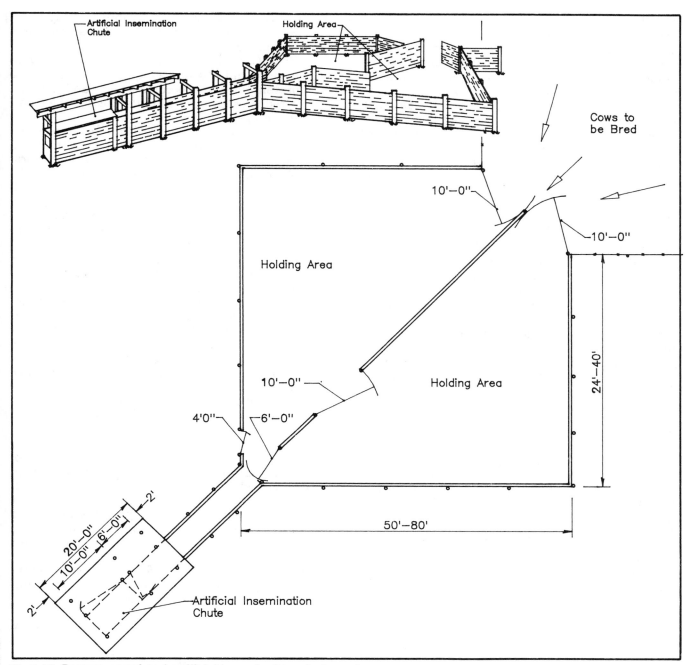

Fig 5-13. Pasture corral and artificial insemination chute system.
Cows to be bred are moved into smaller holding pens and grouped for morning breeding. Stall shown is minimum size. Solid sides on working pens aid cattle movement. USDA plan no. 6152.

this setup, additional stalls can be added without affecting the design.

Bull Pens and Barns

The number of bulls needed depends on the number of cows and heifers, size and age of bulls, crossbreeding program, available pasture, and length of breeding season. Provide space for at least one bull per 25 cows or heifers.

Bull performance and working life are affected by management and housing. Good bull management ensures:

- Bull fertility at the start of every breeding season.
- Good bull health.
- Sound, well trimmed feet.
- Proper feeding.
- Comfortable, clean, dry, and draft-free housing.
- Minimal risk of injury to handlers, bulls, and animals to be bred.

A bull pen must provide a healthy environment which encourages good management with a minimum of risk to both the bull and beef producer.

When possible, pen bulls separately for individual care and to reduce fighting. Up to 10 bulls can be penned together to reduce investment and labor for feeding, watering, cleaning, and care. Provide a handling alley and access to a working chute and headgate-squeeze from the bull pen area.

Protection from subzero winds and weather is essential to prevent injury to testicles from freezing, which can cause sterility. Bulls usually do not need an enclosed barn if correctly fed, and a windbreak and shade are provided. Provide barn space for very cold or hot conditions, grooming, treatment, observation, isolation, and other handling.

Bull Pen Design

Provide at least 200 ft²/bull when housing bulls for less than a week at a time. A pen of this size is adequate for bulls that run with cows for part of the winter and are turned out on pasture over the summer months. Keeping a bull in a small area for too long can lead to lameness and breeding difficulties.

Provide an exercise area in addition to the bull pen for bulls that are housed for most of the year. Sufficient space is needed to keep bulls healthy and in good condition. Provide 1,200 to 1,500 ft²/bull.

A major problem with housed bulls is overgrown feet, which results from insufficient exercise (too small a bull pen) and soft, wet bedding. Design bull pens with hard floors that drain well and require minimal bedding. Clean out pens daily or at least weekly, so bulls are standing on clean dry concrete. Clean, dry bull pens are important to good animal health. Timber railway sleepers can form a raised, dry bed. Insulate below concrete floors to reduce heat loss through the floor.

Safety

Safety is an important factor in bull housing and handling. Bulls can be dangerous animals, so design pens to minimize risk. During planning, assume all bulls are dangerous. Most accidents occur because of poor pen design causing producers to place themselves at risk when handling bulls. Consider the following practices when planning bull housing.

- **Bull restraint.** Never go along side a bull unless it is properly restrained.
- **Visual contact.** Provide an interesting environment for bulls (who are inquisitive by nature) to prevent them from becoming bored and bad-tempered. Design pens so bulls can see their surroundings.
- **Feeding.** Design and position feedbunks and waterers to allow filling and cleaning from outside the pen. Anchor water bowls firmly (preferably bricked up below) to prevent damage to or by the bull. A headgate can be included in the feedbunk design for better restraint. Feeding daily through the headgate ensures that the bull can be easily caught when it has to be handled or moved. A gate next to the feedbunk can swing around and restrain the bull.
- **Safety pass or post.** Provide a safety pass and/or safety post in all bull pens in case of emergencies. Use 14" wide wall or fence openings or vertical steel posts across the corner of the pen. Use wall or fence slits cautiously because they allow young children to enter the pen. With posts in the corner, make the squeeze opening 14" wide. Provide a way for a person to climb out of the pen without having to wait to be rescued. Locate the climbout on the inside of the bull pen to prevent unattended children from climbing into the pen.

Pasture

Graze bulls outside during the summer whenever possible. Grazing bulls with a few dry cows helps control behavior. Two-strand electric fences can control bulls trained to recognize an electric fence before being turned out. Train bulls by setting up a small length of electric fence in the bull pen, so bulls can come in contact with it without breaking through a fence.

Provide a more durable fence in areas near the farmstead and places where the possibility of bulls getting out can cause a dangerous situation.

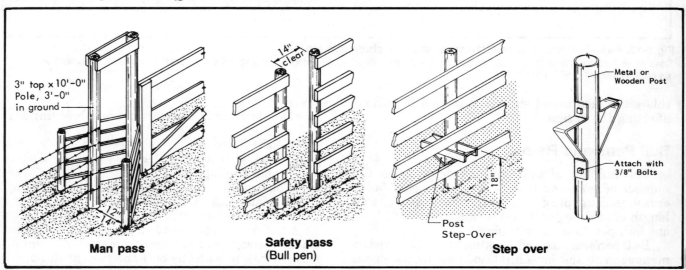

3" top x 10'-0" Pole, 3'-0" in ground

14" clear

18"

Post Step-Over

Metal or Wooden Post

Attach with 3/8" Bolts

12"

14"

Man pass **Safety pass (Bull pen)** **Step over**

Fig 5-14. Safety passes.

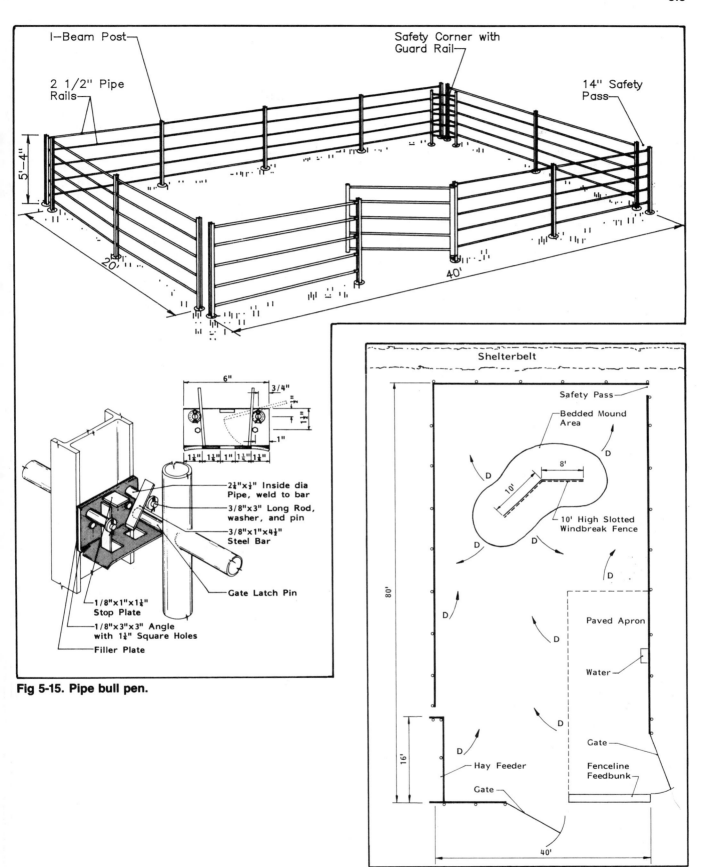

Fig 5-15. Pipe bull pen.

I–Beam Post

2 1/2" Pipe Rails

5'–4"

20'

40'

Safety Corner with Guard Rail

14" Safety Pass

6"

3/4"

1"

2¼"x¼" Inside dia Pipe, weld to bar

3/8"x3" Long Rod, washer, and pin

3/8"x1"x4½" Steel Bar

Gate Latch Pin

1/8"x1"x1¼" Stop Plate

1/8"x3"x3" Angle with 1¼" Square Holes

Filler Plate

Shelterbelt

Safety Pass

Bedded Mound Area

8'

10'

D

D

D

D

D

D

D

D

D

D

10' High Slotted Windbreak Fence

Paved Apron

Water

Gate

Fenceline Feedbunk

Hay Feeder

Gate

80'

16'

40'

Fig 5-16. Bull pen with windbreak.
This layout is adequate for 6 to 8 head.

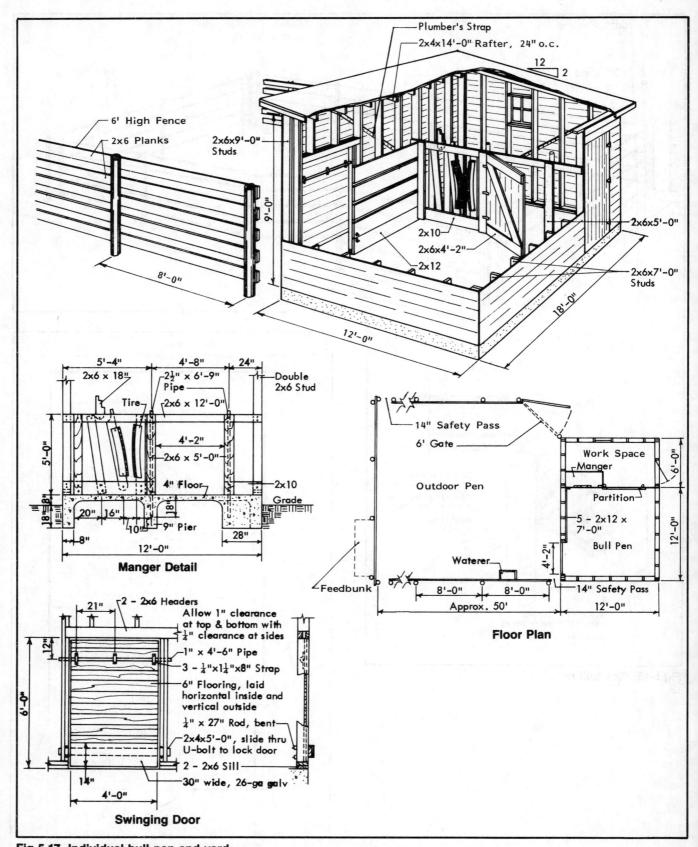

Manger Detail

Swinging Door

Floor Plan

Fig 5-17. Individual bull pen and yard.
Use for special care, observation, protection, etc. A 12'x16' insulated work area is recommended for grooming.

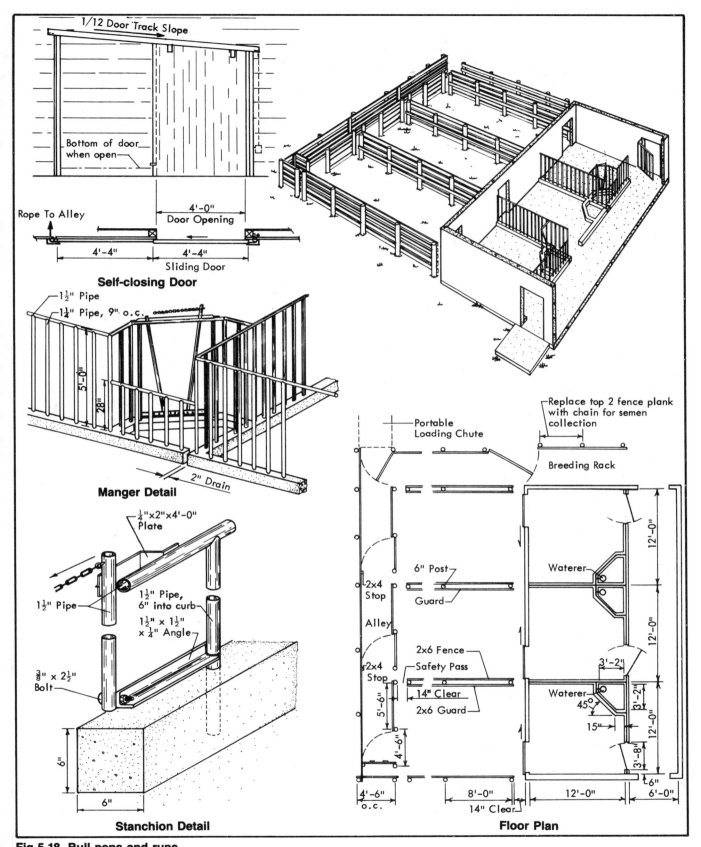

Self-closing Door

1/12 Door Track Slope

Bottom of door when open

Rope To Alley

4'-0"
Door Opening

4'-4"   4'-4"
Sliding Door

Manger Detail

1½" Pipe
1¼" Pipe, 9" o.c.
5'-0"
28"
2" Drain

Stanchion Detail

¼"x2"x4'-0" Plate
1½" Pipe
1½" Pipe, 6" into curb
1½" x 1½" x ¼" Angle
⅜" x 2½" Bolt
6"
6"

Floor Plan

Replace top 2 fence plank with chain for semen collection

Breeding Rack

Portable Loading Chute

6" Post
Guard

2x6 Fence
Safety Pass
14" Clear
2x6 Guard

2x4 Stop
Alley
2x4 Stop
5'-6"
4'-6"

4'-6" o.c.

8'-0"
14" Clear

Waterer
12'-0"
12'-0"
Waterer
45°
15"
3'-2"
3'-2"
3'-1"
3'-8"
12'-0"

12'-0"   6"   6'-0"

Fig 5-18. Bull pens and runs.
Provide larger outdoor pens where snow cleanout is a problem. A south or east exposure is desirable with drainage away from the barn. Size doors to permit skid-steer cleaning.

Cattle Treatment Facility

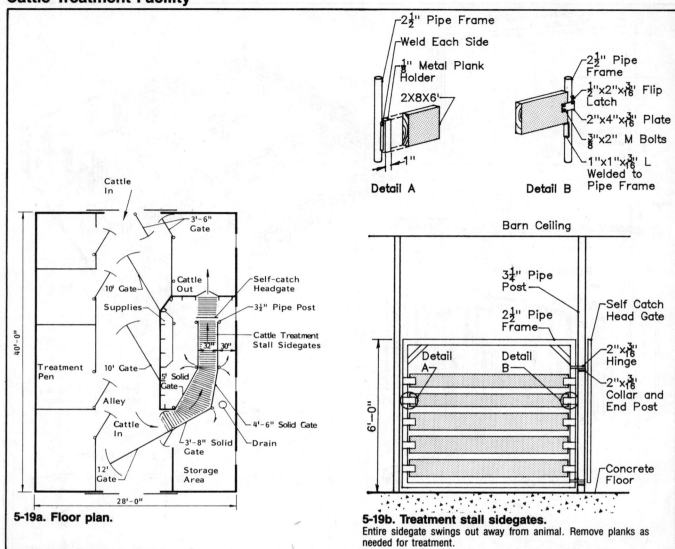

Detail A

- 2½" Pipe Frame
- Weld Each Side
- ⅛" Metal Plank Holder
- 2X8X6'
- 1"

Detail B

- 2½" Pipe Frame
- ½"x2"x³⁄₁₆" Flip Latch
- 2"x4"x³⁄₁₆" Plate
- ⅜"x2" M Bolts
- 1"x1"x³⁄₁₆" L Welded to Pipe Frame

Floor plan (5-19a):
- Cattle In
- 3'-6" Gate
- Cattle Out
- Self-catch Headgate
- 10' Gate
- 3½" Pipe Post
- Supplies
- Cattle Treatment Stall Sidegates
- 32" 30"
- Treatment Pen
- 10' Gate
- 5' Solid Gate
- Alley
- Cattle In
- 4'-6" Solid Gate
- 3'-8" Solid Gate
- Drain
- 12' Gate
- Storage Area
- 40'-0"
- 28'-0"

5-19a. Floor plan.

Treatment stall sidegates (5-19b):
- Barn Ceiling
- 3¼" Pipe Post
- 2½" Pipe Frame
- Self Catch Head Gate
- Detail A
- Detail B
- 2"x³⁄₁₆" Hinge
- 2"x³⁄₁₆" Collar and End Post
- 6'-0"
- Concrete Floor

5-19b. Treatment stall sidegates.
Entire sidegate swings out away from animal. Remove planks as needed for treatment.

Fig 5-19. Treatment facility.

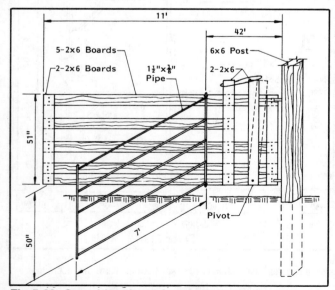

- 11'
- 42'
- 5-2x6 Boards
- 6x6 Post
- 2-2x6 Boards
- 1½"x⅛" Pipe
- 2-2x6
- 51"
- 50"
- 7'
- Pivot

Fig 5-20. Crowd headgate for mothering pen.

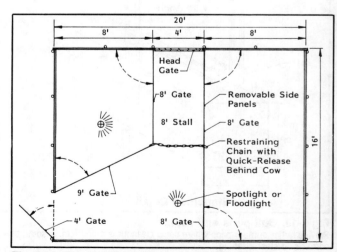

- 20'
- 8' 4' 8'
- Head Gate
- 8' Gate
- Removable Side Panels
- 8' Stall
- 8' Gate
- Restraining Chain with Quick-Release Behind Cow
- 9' Gate
- Spotlight or Floodlight
- 4' Gate
- 8' Gate
- 16'

Fig 5-21. Treatment stall in pen space.

Equipment

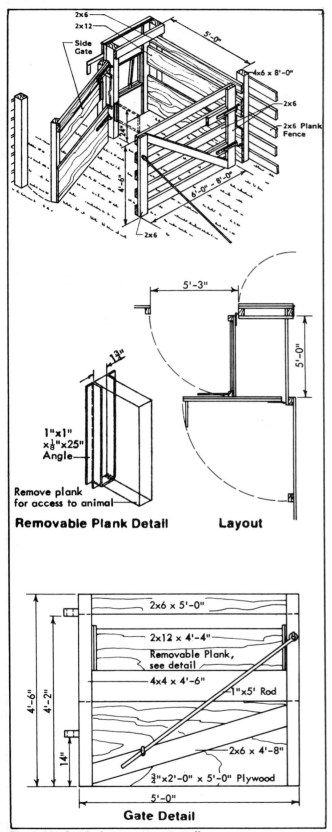

Removable Plank Detail   **Layout**

2x6
2x12
Side Gate
5'-0"
4x6 x 8'-0"
2x6
2x6 Plank Fence
24"
4'-6"
6'-0" - 8'-0"
2x6

5'-3"
5'-0"

1½"

1"x1" x⅛"x25" Angle

Remove plank for access to animal

Gate Detail

2x6 x 5'-0"
2x12 x 4'-4"
Removable Plank, see detail
4x4 x 4'-6"
1"x5' Rod
2x6 x 4'-8"
¾"x2'-0" x 5'-0" Plywood
4'-6"
4'-2"
14"
5'-0"

Fig 5-22. Cattle lot treatment stall.
Stall is suitable for lot corner or hospital pen. For calving, use snub instead of headgate.

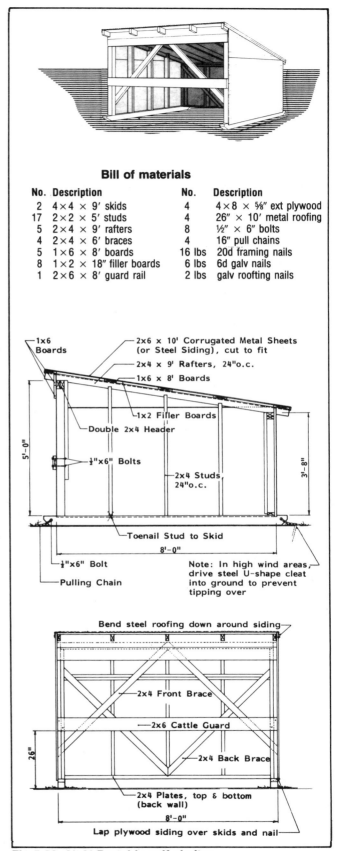

Bill of materials

No.	Description	No.	Description
2	4×4 × 9' skids	4	4×8 × ⅝" ext plywood
17	2×2 × 5' studs	4	26" × 10' metal roofing
5	2×4 × 9' rafters	8	½" × 6" bolts
4	2×4 × 6' braces	4	16" pull chains
5	1×6 × 8' boards	16 lbs	20d framing nails
8	1×2 × 18" filler boards	6 lbs	6d galv nails
1	2×6 × 8' guard rail	2 lbs	galv roofing nails

1x6 Boards
2x6 x 10' Corrugated Metal Sheets (or Steel Siding), cut to fit
2x4 x 9' Rafters, 24"o.c.
1x6 x 8' Boards
1x2 Filler Boards
Double 2x4 Header
½"x6" Bolts
2x4 Studs, 24"o.c.
5'-0"
3'-8"
Toenail Stud to Skid
½"x6" Bolt
8'-0"
Pulling Chain
Note: In high wind areas, drive steel U-shape cleat into ground to prevent tipping over

Bend steel roofing down around siding
2x4 Front Brace
2x6 Cattle Guard
2x4 Back Brace
26"
2x4 Plates, top & bottom (back wall)
8'-0"
Lap plywood siding over skids and nail

Fig 5-23. 8'x8' Portable calf shelter.

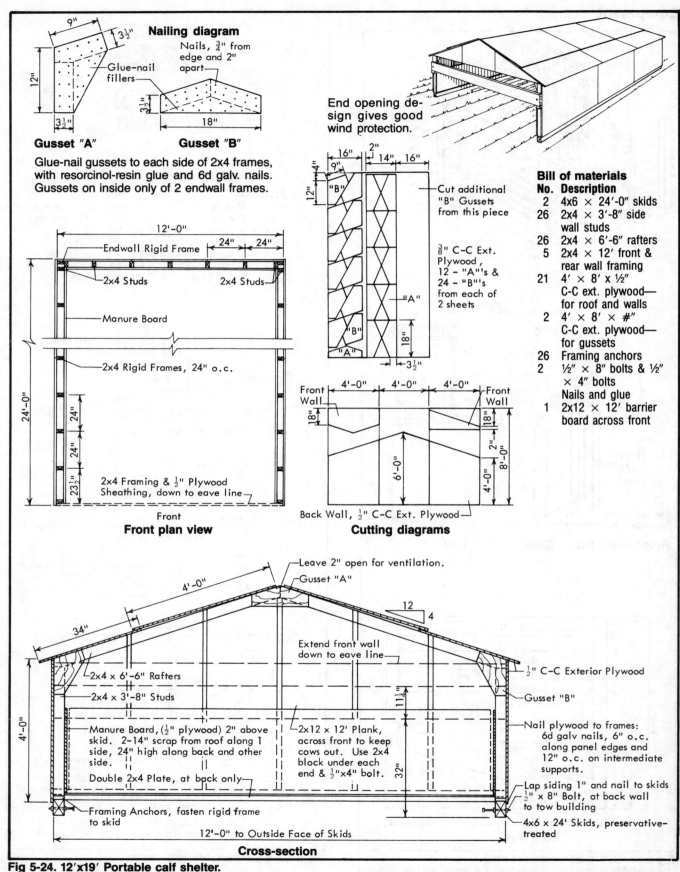

Nailing diagram

Nails, ¾" from edge and 2" apart

Glue-nail fillers

Gusset "A"   9"   3½"   12"   3½"

Gusset "B"   3½"   18"

Glue-nail gussets to each side of 2x4 frames, with resorcinol-resin glue and 6d galv. nails. Gussets on inside only of 2 endwall frames.

End opening design gives good wind protection.

Cut additional "B" Gussets from this piece

⅜" C-C Ext. Plywood, 12 - "A"'s & 24 - "B"'s from each of 2 sheets

16"   2"   14"   16"
4"   9"   "B"   12"   "A"   18"   "B"   "A"   3½"

Bill of materials

No.	Description
2	4x6 × 24'-0" skids
26	2x4 × 3'-8" side wall studs
26	2x4 × 6'-6" rafters
5	2x4 × 12' front & rear wall framing
21	4' × 8' × ½" C-C ext. plywood— for roof and walls
2	4' × 8' × #" C-C ext. plywood— for gussets
26	Framing anchors
2	½" × 8" bolts & ½" × 4" bolts Nails and glue
1	2x12 × 12' barrier board across front

Front plan view

12'-0"   24"   24"

Endwall Rigid Frame

2x4 Studs   2x4 Studs

Manure Board

2x4 Rigid Frames, 24" o.c.

24'-0"   24"   24"   24"   23½"

2x4 Framing & ½" Plywood Sheathing, down to eave line

Front

Cutting diagrams

Front Wall   4'-0"   4'-0"   4'-0"   Front Wall
18"   18"   2"
6'-0"   4'-0"   8'-0"

Back Wall, ½" C-C Ext. Plywood

Cross-section

Leave 2" open for ventilation.

Gusset "A"

4'-0"   12   4

34"

Extend front wall down to eave line

½" C-C Exterior Plywood

Gusset "B"

2x4 × 6'-6" Rafters

2x4 × 3'-8" Studs

11¼"

Manure Board, (½" plywood) 2" above skid. 2-14" scrap from roof along 1 side, 24" high along back and other side.

Double 2x4 Plate, at back only

2x12 × 12' Plank, across front to keep cows out. Use 2x4 block under each end & ½"×4" bolt.

32"

Nail plywood to frames: 6d galv nails, 6" o.c. along panel edges and 12" o.c. on intermediate supports.

4'-0"

Framing Anchors, fasten rigid frame to skid

Lap siding 1" and nail to skids ½" × 8" Bolt, at back wall to tow building

4x6 × 24' Skids, preservative-treated

12'-0" to Outside Face of Skids

Fig 5-24. 12'x19' Portable calf shelter.
For 12 to 15 calves. Bed daily and move frequently to help reduce scours. Locate open side for maximum sun penetration. Anchor corners to prevent wind damage.

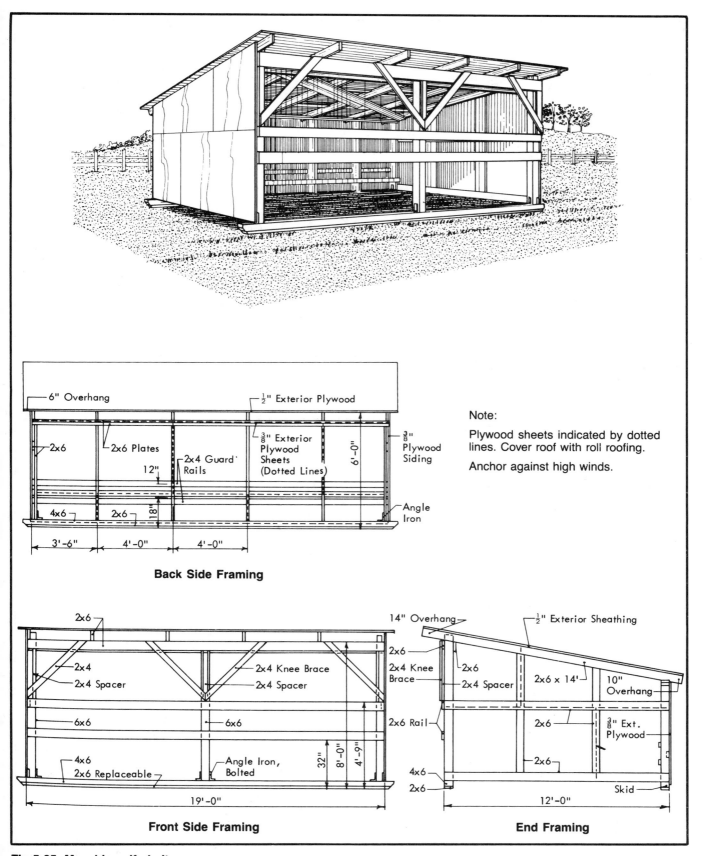

Note:

Plywood sheets indicated by dotted lines. Cover roof with roll roofing.

Anchor against high winds.

Back Side Framing

Front Side Framing

End Framing

Fig 5-25. Movable calf shelter.
Enclosed design gives best wind protection for use on open range. Anchor corners with short posts to prevent wind damage.

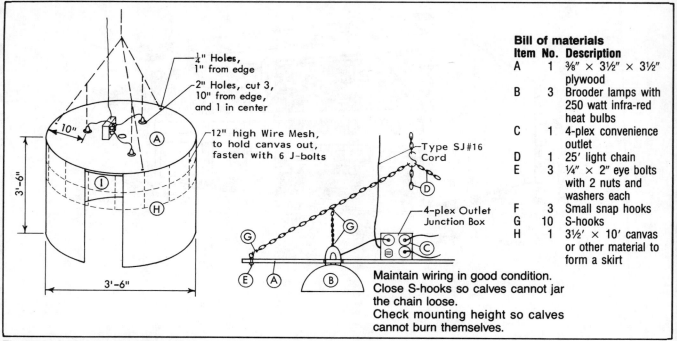

Bill of materials

Item	No.	Description
A	1	⅜" × 3½" × 3½" plywood
B	3	Brooder lamps with 250 watt infra-red heat bulbs
C	1	4-plex convenience outlet
D	1	25' light chain
E	3	¼" × 2" eye bolts with 2 nuts and washers each
F	3	Small snap hooks
G	10	S-hooks
H	1	3½' × 10' canvas or other material to form a skirt

Maintain wiring in good condition.
Close S-hooks so calves cannot jar the chain loose.
Check mounting height so calves cannot burn themselves.

Fig 5-26. Calf pen portable heater.
Suspend from above by S-hooks and a chain. Move heater to pens as needed to warm weak calves. Check mounting height, so calves do not burn themselves.

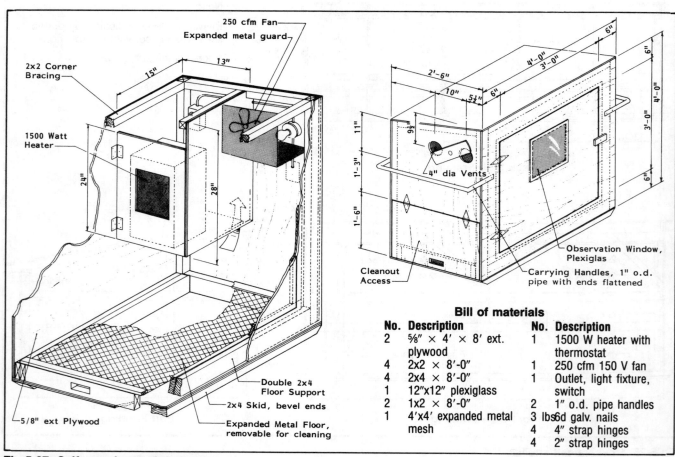

Bill of materials

No.	Description	No.	Description
2	⅝" × 4' × 8' ext. plywood	1	1500 W heater with thermostat
4	2x2 × 8'-0"	1	250 cfm 150 V fan
4	2x4 × 8'-0"	1	Outlet, light fixture, switch
1	12"x12" plexiglass	2	1" o.d. pipe handles
2	1x2 × 8'-0"	3 lbs	6d galv. nails
1	4'x4' expanded metal mesh	4	4" strap hinges
		4	2" strap hinges

Fig 5-27. Calf warming and drying box.
Operate at 90 F. Place newborn calves in unit for about 2 hr to dry off and warm up.

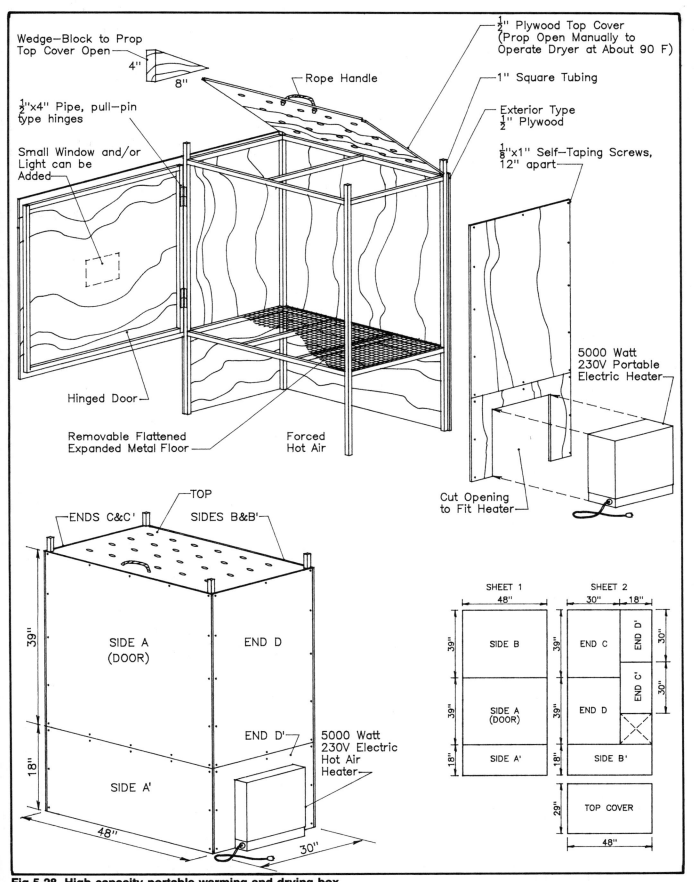

Wedge—Block to Prop Top Cover Open

4"

8"

½"x4" Pipe, pull—pin type hinges

Small Window and/or Light can be Added

Hinged Door

Removable Flattened Expanded Metal Floor

Forced Hot Air

Rope Handle

½" Plywood Top Cover (Prop Open Manually to Operate Dryer at About 90 F)

1" Square Tubing

Exterior Type ½" Plywood

⅛"x1" Self—Taping Screws, 12" apart

5000 Watt 230V Portable Electric Heater

Cut Opening to Fit Heater

TOP

ENDS C&C'

SIDES B&B'

39"

18"

48"

SIDE A (DOOR)

END D

SIDE A'

END D'

5000 Watt 230V Electric Hot Air Heater

30"

SHEET 1

48"

39"

39"

18"

SIDE B

SIDE A (DOOR)

SIDE A'

SHEET 2

30"   18"

39"

18"

END C

END D

SIDE B'

END D'   30"

END C'   30"

29"

TOP COVER

48"

Fig 5-28. High capacity portable warming and drying box.
Warm newborn calves in severe weather. Calves usually dry in less than an hour. Use some bedding to prevent injury and aid cleaning.

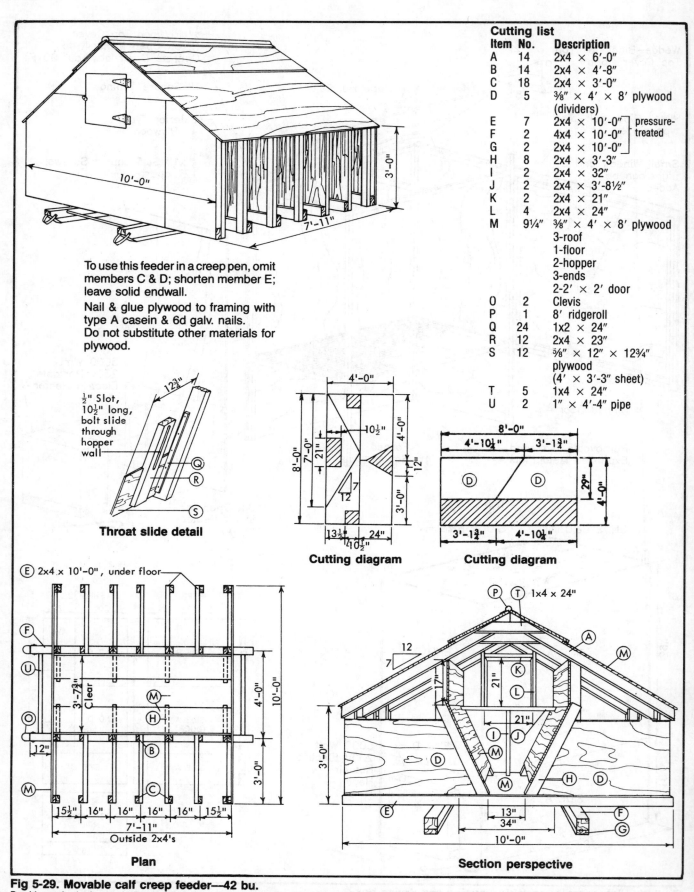

Cutting list

Item	No.	Description
A	14	2x4 × 6'-0"
B	14	2x4 × 4'-8"
C	18	2x4 × 3'-0"
D	5	⅜" × 4' × 8' plywood (dividers)
E	7	2x4 × 10'-0" ⎤ pressure-
F	2	4x4 × 10'-0" ⎬ treated
G	2	2x4 × 10'-0" ⎦
H	8	2x4 × 3'-3"
I	2	2x4 × 32"
J	2	2x4 × 3'-8½"
K	2	2x4 × 21"
L	4	2x4 × 24"
M	9¼"	⅜" × 4' × 8' plywood 3-roof 1-floor 2-hopper 3-ends 2-2' × 2' door
O	2	Clevis
P	1	8' ridgeroll
Q	24	1x2 × 24"
R	12	2x4 × 23"
S	12	⅝" × 12" × 12¾" plywood (4' × 3'-3" sheet)
T	5	1x4 × 24"
U	2	1" × 4'-4" pipe

To use this feeder in a creep pen, omit members C & D; shorten member E; leave solid endwall.

Nail & glue plywood to framing with type A casein & 6d galv. nails.

Do not substitute other materials for plywood.

½" Slot, 10½" long, bolt slide through hopper wall

Throat slide detail

Cutting diagram

Cutting diagram

Ⓔ 2x4 x 10'-0", under floor

Plan

Section perspective

Fig 5-29. Movable calf creep feeder—42 bu.
Provide one feeder for every 40 to 50 calves. Stall partitions allow timid calves to eat with other more aggressive calves. The individual partitions can cause turning problems—calves can get caught.

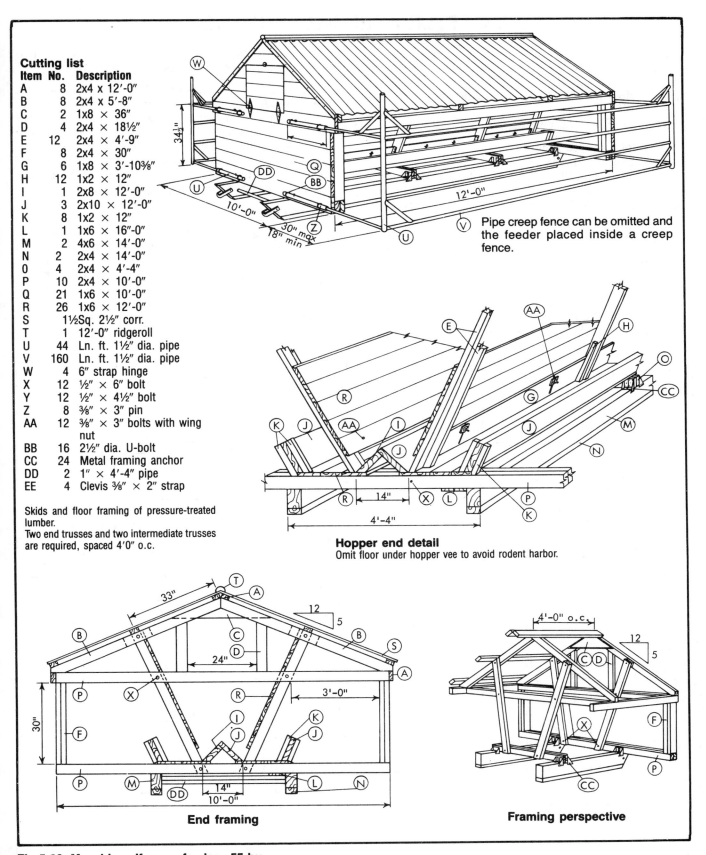

Cutting list

Item	No.	Description
A	8	2x4 x 12'-0"
B	8	2x4 x 5'-8"
C	2	1x8 × 36"
D	4	2x4 × 18½"
E	12	2x4 × 4'-9"
F	8	2x4 × 30"
G	6	1x8 × 3'-10⅜"
H	12	1x2 × 12"
I	1	2x8 × 12'-0"
J	3	2x10 × 12'-0"
K	8	1x2 × 12"
L	1	1x6 × 16'-0"
M	2	4x6 × 14'-0"
N	2	2x4 × 14'-0"
O	4	2x4 × 4'-4"
P	10	2x4 × 10'-0"
Q	21	1x6 × 10'-0"
R	26	1x6 × 12'-0"
S		1½Sq. 2½" corr.
T	1	12'-0" ridgeroll
U	44	Ln. ft. 1½" dia. pipe
V	160	Ln. ft. 1½" dia. pipe
W	4	6" strap hinge
X	12	½" × 6" bolt
Y	12	½" × 4½" bolt
Z	8	⅜" × 3" pin
AA	12	⅜" × 3" bolts with wing nut
BB	16	2½" dia. U-bolt
CC	24	Metal framing anchor
DD	2	1" × 4'-4" pipe
EE	4	Clevis ⅜" × 2" strap

Skids and floor framing of pressure-treated lumber.

Two end trusses and two intermediate trusses are required, spaced 4'0" o.c.

V Pipe creep fence can be omitted and the feeder placed inside a creep fence.

Hopper end detail
Omit floor under hopper vee to avoid rodent harbor.

End framing

Framing perspective

Fig 5-30. Movable calf creep feeder—55 bu.
Provide one feeder for every 65 calves.

6. CATTLE FEEDING FACILITIES

Cattle feeding systems vary with size of operation, drainage, climate, ration, and available facilities. Cattle feeding facilities are outdoor feedlots, barns with outdoor pens, or covered confinement systems.

Outdoor Feedlot Systems

Feedlot features include:
- Fenceline bunks.
- Lot sizes of 120 to 240 head. Lots usually hold multiples of an average truckload of 60 head.
- Lots arranged around a central office area.
- Feeding pens near the feed mill to reduce travel.
- Traffic lanes and feedbunks oriented north-south to maximize snow melting and drying.
- Lot corners rounded at intersections for easier turning of cattle and vehicles.
- A service area, separate receiving and load-out areas, cattle working facilities, separate conditioning lots, and hospital facilities.
- Separate vehicle and cattle traffic lanes when possible to minimize congestion and reduce spread of parasites, disease, and manure.

Feedlot area depends on cattle numbers and size, space needed for manure and snow storage, and equipment use. As a rule of thumb, plan for 100 head/acre. Space needs vary with the amount of paved space, soil type, drainage, annual rainfall, and freezing and thawing cycles. Specific needs depend on climate and drainage. See the planning data summary, Table 1-1, for recommended lot areas.

Open Feedlot

Open feedlots are usually unpaved. Mounds improve drainage and provide areas that dry quickly; a dry resting area improves cattle comfort, health, and feed utilization. With open feedlots, weather protection is often limited to a windbreak fence in the winter and/or a sunshade in the summer. Cattle treatment, working, and hospital areas can be covered.

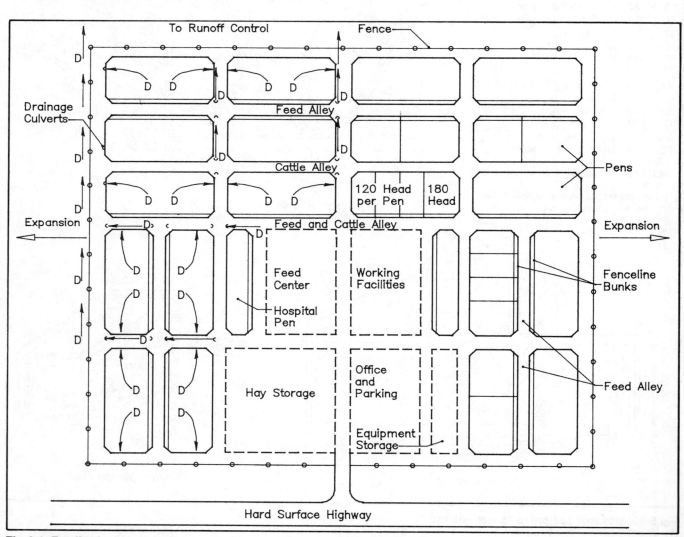

Fig 6-1. Feedlot layout.

Make vehicle drives at least 14' wide and cattle alleys 12' wide. If moving cattle with horses, use 14'-16' wide alleys. When possible, keep vehicle and cattle traffic separate to simplify work routines and reduce tracking manure to feed storage areas. In heavy snowfall or high rainfall areas, a 30'-60' wide drive between two parallel feedbunks provides drainage and snow storage space. Pave all heavy traffic areas. Provide a 10'-12' wide apron along feedbunks and around waterers. All-weather access is important.

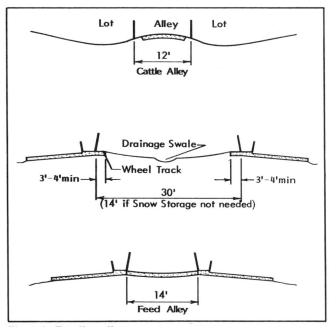

Fig 6-2. Feedlot alleys.

Feedlot planning

Consider lot slope, drainage, and soil type when selecting a feedlot site. Recommended slopes are 4%-6% for good drainage. Slopes greater than 8% can cause excessive erosion. Grade the site to establish the proper slope, if necessary.

Lot space can be affected by lot slope. Minimum requirements are 150 to 250 ft²/hd with 4% or greater slopes, 250 to 400 ft² with 2%-4% lot slopes, and 400 to 800 ft² with slopes below 2%.

Naturally deposited soils, such as shale and impervious soil types, are desirable. With heavy clay soils, develop a manure pack to shed surface water. Do not use rock surfaces because of equipment damage and rock spreading during manure application.

With a sloping site, take advantage of the natural slope to plan drainage and equipment placement. With a flat site, the topography must be shaped. Move earth from the parts designed for settling and runoff control, and build up along fencelines and feedbunks. Additional soil may be needed to achieve desired slopes and drainage. Plan drainage away from feedbunks, waterers, and fencelines.

Provide gates at the end of the bunks for easier pen access with cleaning equipment, Fig 6-3. Gates at the bunk ends make it easier to move cattle between pens but require stopping and starting feeding equipment. For better cattle movement, position gates to open in the direction of most frequent cattle movement. At the lower end of the pen, an angle fence and offset gate provide easier access for moving cattle in and out of the pen.

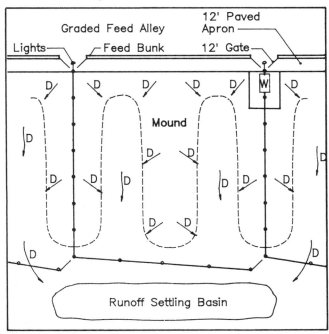

Fig 6-3. Open feedlot layout.
Layout for mild climate. Build up mounds with soil moved for drainage. Provide a shade or shelter for hot weather and cold winds. "D" on an arrow indicates drainage.

Feedlot lighting

In open feedlots, electrical power is needed for lighting and some waterer heaters. Mount overhead wiring and light fixtures 18' or higher and away from bunks and waterers to control bird droppings and insects. Underground wiring is recommended to reduce line failure under ice and snow loads, maintenance, and bird roosting sites. Install wiring according to code. See the utilities chapter for more information.

Barn with Outdoor Pens

Open-front barns and lots with mechanical conveyors or fenceline bunks are common for feedlots up to 1,000 head, especially in areas with severe winter weather and high rainfall. An adjacent outdoor lot is for resting, manure storage, and exercise. When feeding inside a barn, feed and equipment are protected from wind, rain, and snow, and cattle feed intake is more consistent, especially during severe weather. Fig 6-5 shows two barn and lot systems.

Orient open-front barns with the long, open side away from prevailing winter winds. An east or south orientation is usually best. Provide panels in the closed walls for summer ventilation. Connect a paved apron along the open side of the barn to the paved feedbunk apron. Extend aprons 4'-6' into buildings or under roof overhangs to minimize mud caused by

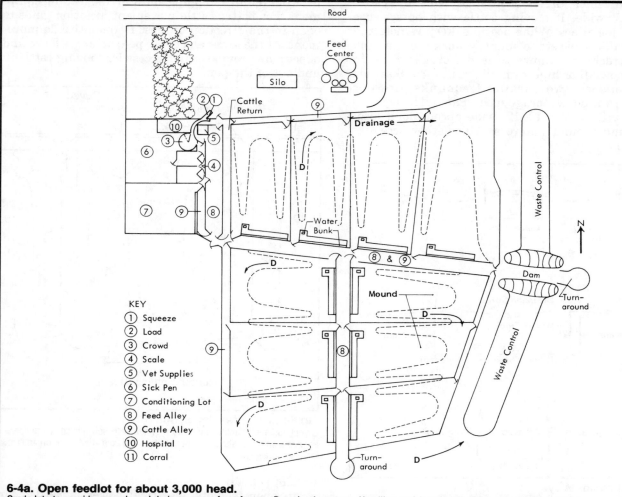

6-4a. Open feedlot for about 3,000 head.
Grade lots to provide mounds and drainage away from fences. Pave bunk aprons. Handling and hospital facilities are combined in this layout. For custom and larger operations, separate handling and hospital facilities.

KEY
1. Squeeze
2. Load
3. Crowd
4. Scale
5. Vet Supplies
6. Sick Pen
7. Conditioning Lot
8. Feed Alley
9. Cattle Alley
10. Hospital
11. Corral

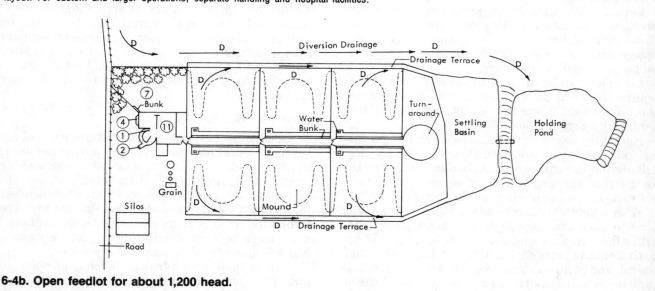

6-4b. Open feedlot for about 1,200 head.

Fig 6-4. Open feedlot systems.
These layouts use drive-through feeding, a central cattle handling facility, feed storage and handling, and a runoff control system.

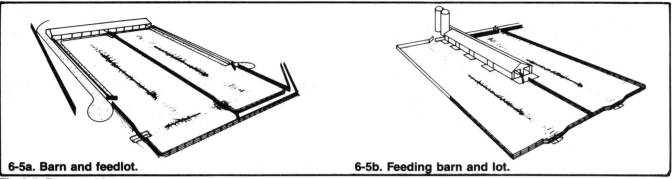

6-5a. Barn and feedlot.   **6-5b. Feeding barn and lot.**

Fig 6-5. Barn and lot system.
Suitable for severe climates.

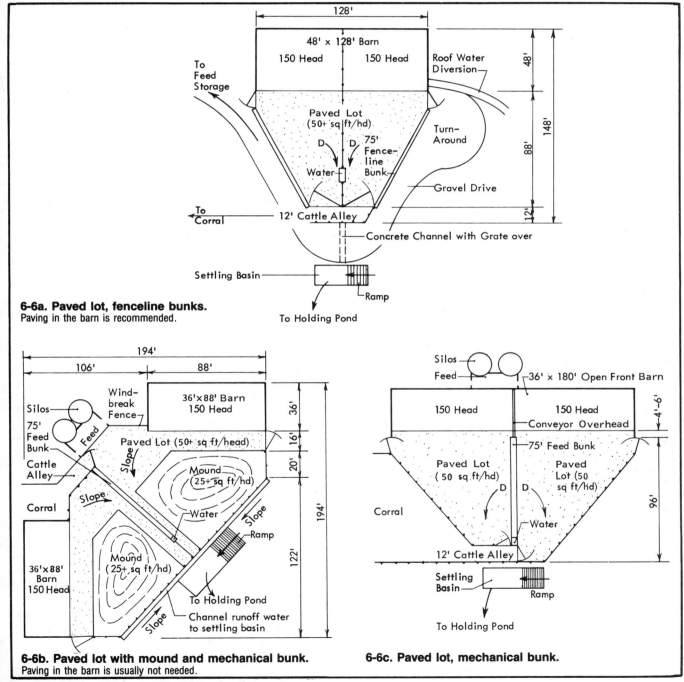

6-6a. Paved lot, fenceline bunks.
Paving in the barn is recommended.

6-6b. Paved lot with mound and mechanical bunk.
Paving in the barn is usually not needed.

6-6c. Paved lot, mechanical bunk.

Fig 6-6. 300-head barn and outdoor feedlot systems.

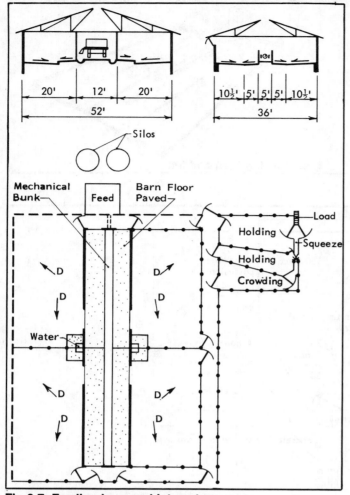

Fig 6-7. Feeding barn and lot system.
Permits indoor feeding so cattle eat regardless of the weather.

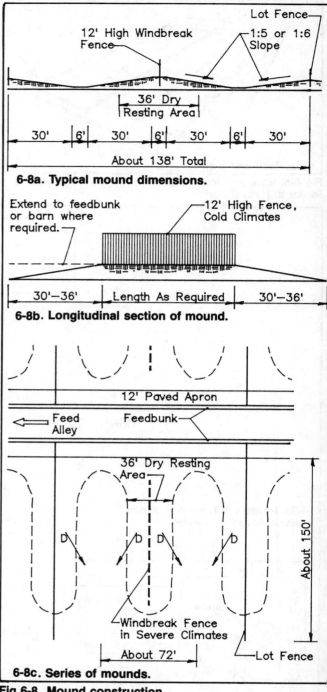

6-8a. Typical mound dimensions.

6-8b. Longitudinal section of mound.

6-8c. Series of mounds.

Fig 6-8. Mound construction.

blown-in snow and rain. Inside barns, make unpaved floors at least 12″, and paved floors at least 6″, above outside grade. Slope floors toward the open side.

Mounds

Orient mounds parallel to the lot slope so runoff does not pond. Begin mounds at feedbunks and slope them away. Round mound tops to avoid holes and promote drainage. At least yearly maintenance is required. A shallow "W" lot cross-sectional shape, Fig 6-8a, is preferred to reduce slope lengths, increase drainage, and reduce manure accumulation along fences.

Stabilize the upper half of the mound surface with chopped bedding, a soil-manure mixture, or by disking in lime. Barn lime reacts more quickly than agricultural limestone. Start by adding 1 lb/ft²; add more lime until mound surface is stable. Power plant fly ash can also stabilize the soil. Fly ash quality varies between power plants and between loads from the same power plant. Be careful to get high quality material; low quality fly ash has poor stabilizing capacity.

A 12′ high, slotted windbreak fence on the mound ridge is recommended in the northern Midwest. The fence provides protection for cattle on one side or the other depending on weather conditions. Provide minimum mound space on each side of the windbreak. Slope lots away from fences to control manure accumulation and insect breeding.

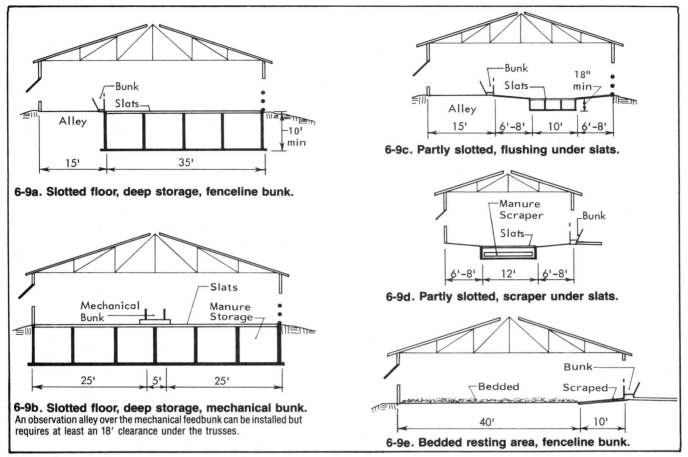

6-9a. Slotted floor, deep storage, fenceline bunk.

6-9c. Partly slotted, flushing under slats.

6-9d. Partly slotted, scraper under slats.

6-9b. Slotted floor, deep storage, mechanical bunk.
An observation alley over the mechanical feedbunk can be installed but requires at least an 18′ clearance under the trusses.

6-9e. Bedded resting area, fenceline bunk.

Fig 6-9. Confinement barn cross sections.
See data summary for space recommendations.

Confinement Feeding Barns

Confinement barns require less land area and solve feedlot problems caused by drifting snow, severe wind, mud, lot runoff, and mound maintenance. A cold confinement barn is usually open to the south or east, depending on prevailing winter winds. Feeding is mechanical or fenceline bunks. Manure systems are bedded-pack, semi-solid, slurry, or liquid. See the manure management chapter.

Provide a well managed natural ventilating system in cold barns to control moisture, modify temperature, and remove gases, odors, and pathogens. See the ventilation chapter. Winter temperatures inside the barn are usually 5 to 10 F above outdoors. Install circulating, heated, or well insulated frost-proof waterers. Birds, flies, and frozen manure can be problems.

Pen size depends on building dimensions, management, and number of cattle purchased or sold at one time. Maximum pen capacity is usually 120 head. Plan pens to group uniform weight cattle. Provide at least 18 ft²/head in slotted floor pens and 30 ft²/head in solid floor pens for cattle 800 to 1,000 lb.

Design bunks for 2 ft³ of feed/ft of bunk length per day with once-per-day feeding. Provide 4″-6″ of bunk space/animal if feed is available at all times. Provide 20″-26″ of bunk space/animal if fed only once a day. See the data summary chapter.

Wall heights are usually 12′-16′ for wagon feeding. Higher walls improve summer comfort but are not a substitute for a well designed and properly managed ventilating system.

Confinement barns can also be completely enclosed, insulated, and mechanically ventilated. These are more expensive to construct and operate, but do eliminate manure freezing and bird problems. Proper design, careful construction, and good management are essential to a successful operation.

Floors

Slotted floors require less space/head, reduce daily manure removal labor, and keep cattle cleaner than solid floors. Concrete slats are typical for beef buildings. Slat spacing is usually 1¼″-1¾″ and slats are up to 12″ wide. Narrow slats (6″-8″) result in cleaner cattle. Locate waterers across the pen from bunks to increase animal traffic, improve manure passage through slotted floors, and reduce the amount of feed deposited in waterers.

7. CATTLE HANDLING FACILITIES

Cattle handling includes sorting, weighing, dehorning, vaccinating, dipping, branding, grooming, treatment, and calving. Well designed cattle handling facilities minimize labor and allow for safe cattle handling.

The components of a facility are the same regardless of the number of cattle and include:
- Conditioning, holding, sorting, and crowding pens.
- Working and loading chutes.
- A squeeze and headgate.
- Dipping facilities.
- A weigh scale.
- Hospital facilities.

Differences in size and number of pens, working chute, type of headgate, etc., depend on cattle size and number. Locate handling facilities close to cattle pens and yard area for easy access. Provide 300' or more between residences and handling facilities to reduce noise and dust. Select a well drained site with an all-weather access road. An indoor working chute-headgate allows working in all types of weather at any time of day. Slope working facilities less than 3% to reduce gate swing problems.

Conditioning Pens

Conditioning (or receiving) pens hold incoming cattle before they go to the feedlot. New cattle are usually stressed from weaning, removal from range, crowding in trucks and rail cars, motion sickness, thirst, hunger, and fright. Eating from a bunk or drinking from a tank or float controlled waterer may be a new experience.

To help calves adapt from pasture to confinement:
- Do not crowd animals. Provide well drained conditioning pens with 100 ft²/hd in lots that provide good footing. Avoid slippery surfaces and slotted floors. Limit pens to 60 hd/pen. Large pens encourage running, and may cause dust, trouble finding feed or water, and additional stress.
- Fence visibility is important in conditioning pens. Plank fences are easily seen by calves; wire and cable fences are more difficult to see. Do not use wire and cable fences unless at least one plank is attached at calf eye level.
- Provide wind protection and shelter from sun, rain, snow, etc. Protected feedbunks help maintain feed quality and uniform consumption.
- Provide at least 2' of bunk space per hd. Offer hay for 8 to 10 hr before starting other roughages. Fill feeders so cattle can see hay. Start grain or silage gradually during the conditioning period.
- Provide plenty of fresh water. Be careful that calves do not over-water at first by feeding roughage before turning on waterers. Use a running hose because the sound of running water helps new cattle find the waterer.
- Locate receiving lots away from main lots to reduce disease and parasite transmission.

- Process new cattle after a rest period (usually several hours).

Working Facilities

Working facilities are for sorting, handling, and treating cattle. They include the holding pen, crowding pen, working chute, squeeze and headgate, scale, and possibly a dipping or spraying facility.

Holding or Sorting Pen

Make the holding pen about 600 ft² (12'x50', 20'x30', etc) to hold 40 to 50 animals ahead of the crowding pen. One pen is sufficient for operations with up to 250 cattle. For larger operations provide:
- Up to 1,000 hd, one 600 ft² pen/250 head.
- Over 1,000 hd, four 600 ft² pens and two 1000 ft² pens.

Where cattle are moved by persons on foot, provide safety posts in each corner and at 40'-50' intervals along the sides of large pens, Fig 7-1. In large pens, locate safety posts every 50'. Position posts 3'-4' from fences or corners. Use at least 6" posts set 4' in the ground. Safety posts are essential when handling bulls.

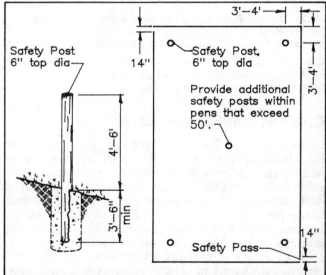

Fig 7-1. Safety post placement.

Crowding or Funnel Pen

A crowding or funnel pen with a swinging gate is needed to crowd cattle into the working chute. Taper the pen from 12' to about 2' at the chute entrance. A circular crowding pen with a solid fence and gate is most effective because the only escape route the cattle can see is through the working chute. Equip all gates with self-locking latches and provide a safety exit from the crowding area.

Working Chute

Desirable characteristics are:
- Curved chute construction with solid sloped

sides that restrict cattle vision to a few feet straight ahead. A minimum curve radius of 15' is recommended.
• Sloped chute sides that restrict an animal's feet to a narrow path and prevent turning around.

Working Facility Layouts

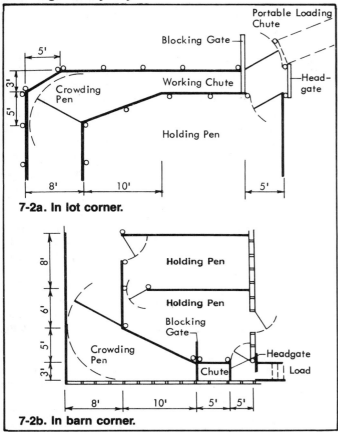

7-2a. In lot corner.

7-2b. In barn corner.

Fig 7-2. Cattle handling layouts for up to about 75 head.

Sloped sides permit working different sized animals in the same chute.
• An overhead restrainer to keep cattle from rearing up and turning around or falling over backward.
• At least a 20' long chute to hold 3 or 4 head at one time. One person working the crowding pen can keep the chute charged to reduce delays at the squeeze and headgate.
• One or two blocking gates to keep cattle from moving forward or backing up. Use near the scale or cutting gates.
• A cutting gate at either the beginning of a chute, or just ahead of the squeeze, or at both places to divert cattle not requiring treatment. Cutting gates are better than running all cattle through the squeeze and headgate.
• A concrete floor or other all-weather surface for easier cleaning. Use a sloped (less than 4%), rough finished surface for good traction. See the building construction chapter for slip resistant floor surfaces.
• Chute sides about 4" above ground to improve manure removal and control insect breeding.

Squeeze and Headgate

A squeeze and headgate restrains cattle for treatment. Usually a headgate on a stall with fixed sides is satisfactory for small cow-calf operations. A squeeze provides faster, more complete animal control, reducing the chance of injury to animal or operator, Fig 7-2.

Consider a tilting table in the squeeze-headgate area for branding, castrating, hoof trimming, and treatment. Select a headgate which opens the full width of the chute to reduce cattle injuries. Size the chute and headgate based on cattle being worked, Table 7-1.

Three headgate types are:
• **A self-catch gate** is easiest for one person to

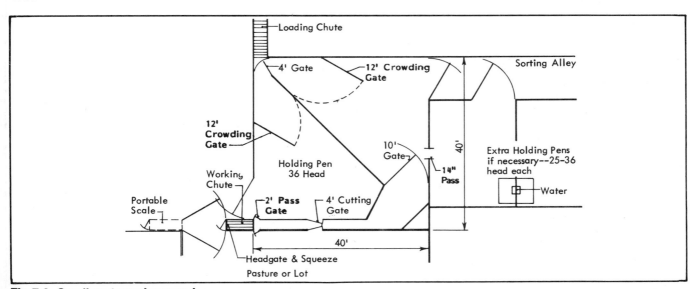

Fig 7-3. Small rectangular corral.
For small feedyards or cow-calf operations. This layout is simple and easier to construct but has lower capacity.

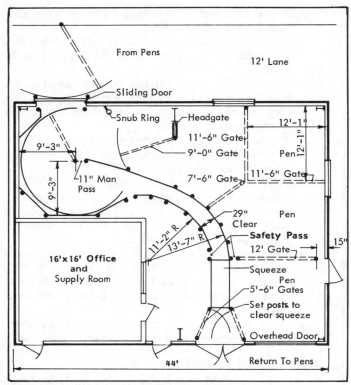

Fig 7-4. Cattle handling facility.
Adaptable as a hospital barn. Gates swing out of the way for tractor cleaning. Mount overhead lighting in the working area to see head, sides, and tail of animals in squeeze or headgate and in the corner pen for care during calving.

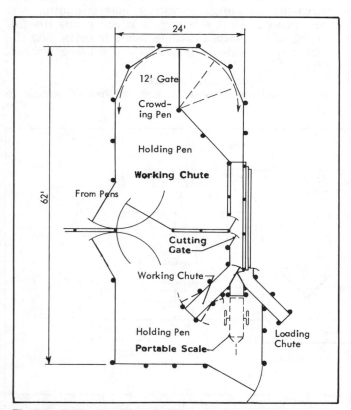

Fig 7-5. Cattle handling facilities for 300 to 1,000 head.

operate; cattle work fast, exit easily, and seldom choke. It is difficult to use for small or horned cattle and can cause severe shoulder bruises. Sometimes animals escape without being caught.

- **A stanchion headgate** is lower cost, simple, requires an operator, and seldom chokes cattle. It can cause shoulder bruises and sometimes allows an animal to escape without being caught. Cattle often trip as they walk through the headgate.
- **A guillotine headgate** holds the animal's head down, providing maximum head control and reducing shoulder bruises. This headgate is medium cost, difficult to operate, slower, and of the three types is most likely to cause choking.

The self-catch and stanchion headgates are available in straight bar and curved bar models. A straight bar provides less head control and decreases chances of choking. A curved bar gives better head control, but increases the possibility of choking.

Loading Chute

Well designed loading facilities reduce loading time and injury to animals and operator. Ramp dimensions vary with equipment used. Low bed trailers, less than 12" from the ground, require only a drive alley and side gates to direct cattle, Fig 7-9. Some trucks require wide platforms because of built-in unloading ramps. Provide combination swinging and telescoping gates to close off areas between loading chute sides and truck.

Desired characteristics are:

- A curved approach (30" wide, maximum) with solid sides to keep cattle from seeing operators or trucks until just before loading.
- Telescoping side panels to close the gap between the truck and loading chute.

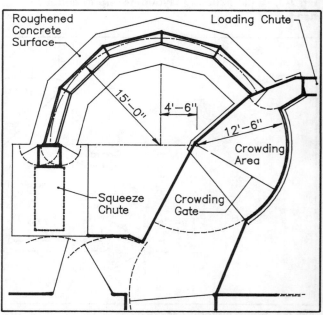

Fig 7-6. Circular crowd pen and working chute.

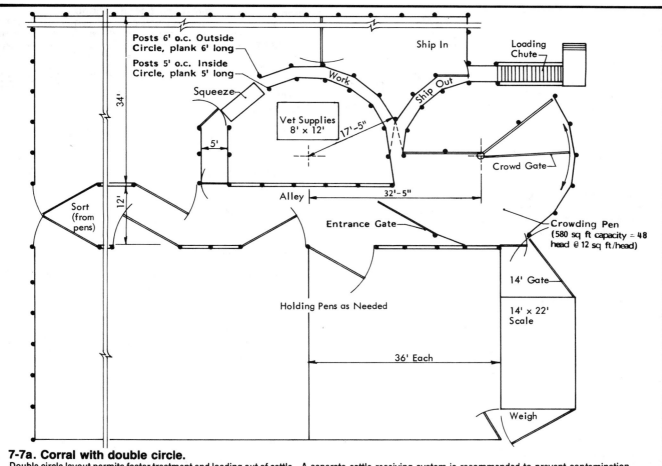

7-7a. Corral with double circle.
Double circle layout permits faster treatment and loading out of cattle. A separate cattle receiving system is recommended to prevent contamination. Refer to Table 7-1 for working chute and fence dimensions.

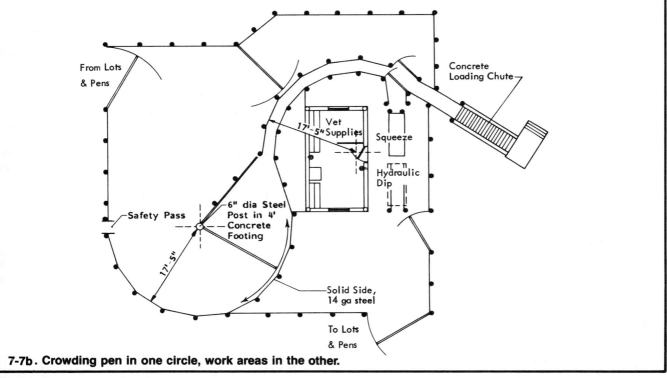

7-7b. Crowding pen in one circle, work areas in the other.

Fig 7-7. Double-circle layout.
Crowding pen, loading chute, and squeeze are combined into an efficient unit.

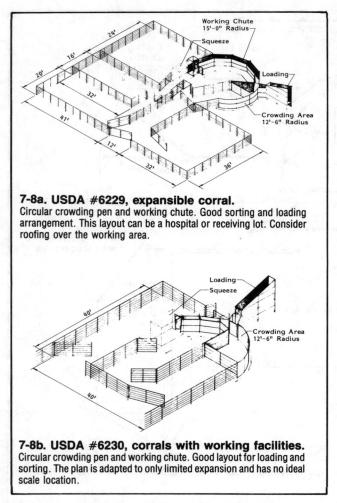

7-8a. USDA #6229, expansible corral.
Circular crowding pen and working chute. Good sorting and loading arrangement. This layout can be a hospital or receiving lot. Consider roofing over the working area.

7-8b. USDA #6230, corrals with working facilities.
Circular crowding pen and working chute. Good layout for loading and sorting. The plan is adapted to only limited expansion and has no ideal scale location.

Fig 7-8. USDA corral plans 6229 and 6230.
These plans show pipe, post, and rail fences. Order plans from the addresses listed on the inside front cover of this book.

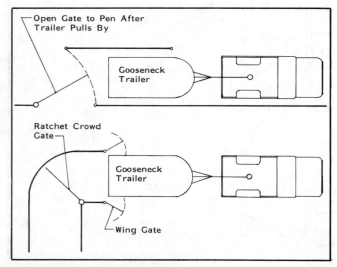

Fig 7-9. Low bed trailer loading.

Table 7-1. Chute and fence specifications.
Set 5″ top pressure preservative treated posts 36″ deep (minimum) in firm ground or concrete for working chute. Leave 7′ minimum headroom below any cross ties. Set 2″ diameter or large steel pipe 30″ deep in concrete backfill.

	To 600 lb	800- 1,200 lb	Over 1,200 lb (or cow-calf)
Holding area, ft²/head	14	17	20
Crowding pen, ft²/head	6	10	12
Working chute, vertical sides			
Width	18″	26″	28″-30″
Desirable length (min.)	20′	20′	20′
Working chute, sloping sides			
Clear width at bottom	13″	13″-16″	18″-20″
Clear width at 4′ height	20″	26″-29″	30″-33″
Minimum length	18′	20′	22′
Working chute fence			
Height—solid lower wall (includes 2″-4 clear at bottom for drainage)	48″	50″	50″
Overall height			
Top rail, quiet cattle	55″	60″	60″
Top rail, range cattle	68″	72″	72″
Corral line fence			
Recommended height	60″	60″	60″
Depth in ground			
5″ top post	36″-42″	36″-42″	36″-42″
2″ steel pipe post	30″	30″	30″
Loading chute			
Width	26″	26″	26″-30″
Length (min.)	12′	12′	12′
Rise, in/ft	3½	3½	3½
Ramp height for:			
Stock trailer - 15″			
Pickup truck - 28″			
Stock truck - 40″			
Tractor-trailer - 48″			
Double deck trailer - 100″			

- A self-aligning dock platform to eliminate the possibility of a gap in the floor between truck and loading chute.
- Gradual slope with cleats or steps. Slope stationary chutes 2½″/ft and adjustable chutes 3″/ft. Make steps 12″ deep and 3½″-4″ high. Use solid material—hollow wood or steel ramps echo, scaring cattle.
- The loading chute located ahead of the curved working chute and using the same crowding pen.
- All-weather access.
- Catwalks with rails for workers.

Dipping Vats

Dipping vats control external parasites. Do not dip cattle immediately after they get off the truck, because they are hot, thirsty, and hungry. Allow cattle to cool off, eat, and drink for several hours before dipping. Mobile swim-through, built-in swim-through, and cage dip vats are common.

Mobile swim-through vat

Mobile vats allow for use at more than one site, ease of changing layout and handling patterns, and a high cattle throughput rate. They can serve any size

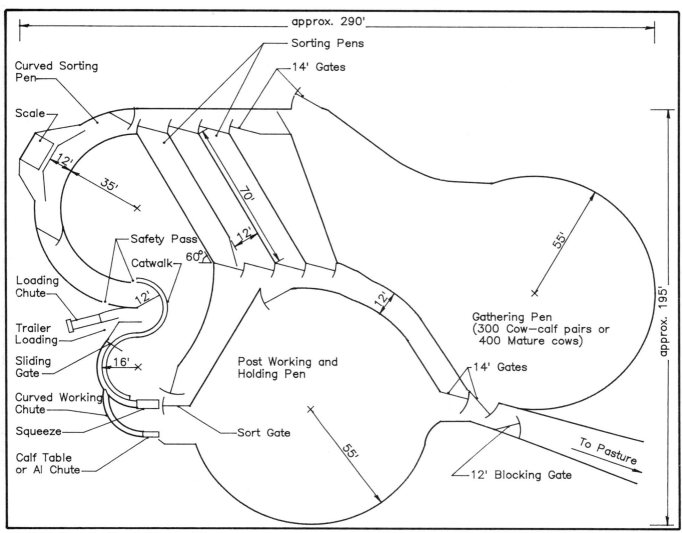

Fig 7-10. Corral with working facility.

operation, especially those too small for a built-in vat. A mobile vat requires maneuvering space, time to move and set up, and portable fence panels and gates.

Cage dip vat

A cage vat is more suitable for operations of less than 10,000 head. Three people can dip up to 100 hd/hr, one animal at a time. A 3'x5'x7' steel cage is hydraulically lowered into the vat. The vat is usually at the end of a working chute. Provide a backup system to rescue an animal if the power fails while the cage is submerged.

Built-in swim-through vat

A built-in swim-through vat is best for large cattle numbers. They are usually installed only at large feedlots because a large quantity of dip solution is required. An experienced crew can dip 500 to 600 hd/hr. Design facilities for both animal and operator safety. Include a method of draining and disposing of used liquid and dirt.

A well designed entrance prevents leaping and keeps cattle from jumping on the backs of others in the vat. A blocking gate and anti-bunch side gates

meter cattle into the vat. An adjustable hold-down device parallel to the lead-in ramp prevents jumping and forces cattle to keep their heads down.

Use a solid lead-in ramp sloped about 25° with cleats for sure footing. Slope the drop off ramp about 45°, Fig 7-12. Sure footing is important because animals panic when they slip on smooth slide-type entrances. Provide a stepped exit ramp, because cattle are more secure on a flat surface than a sloped one. Protect workers with splash shields at the entrance and exit.

After cattle leave the vat, hold them in a drip pen to drip dry and collect carry-out dip solution. Slope the drip pen ¼"/ft to a collection sump, through a filter screen, and back into the vat. Deeply score the concrete floor of the drip pen to prevent slipping.

Use at least 3,500 psi concrete to construct permanent swim-through vats. Make walls at least 8" thick with #4 rebars. Watertight joints are required. Vats are usually 11' deep and 24"-40" wide. Make straight wall vats 36" wide and taper sloped wall vats from 24" at the bottom to 40" at the top, Fig 7-13. Vertical walls are easier to construct, but require more dip solution

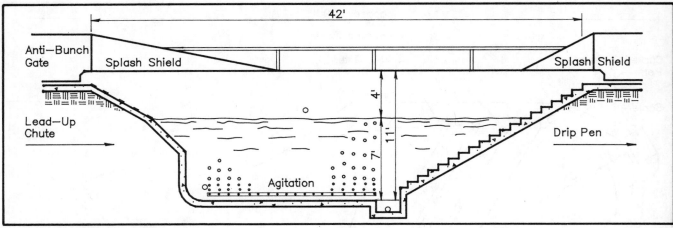

Fig 7-11. Built-in swim-through vat.

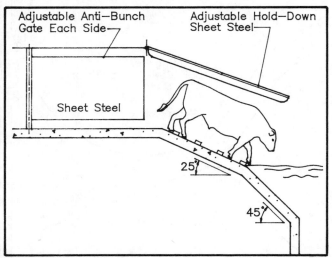

Fig 7-12. Lead-in ramp.

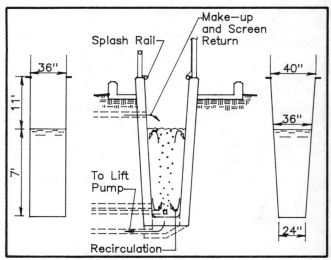

Fig 7-13. Vat walls.

than sloping wall vats. With sloping sidewall vats, fluid level is important. Low fluid levels can allow over-sized cattle to become wedged and drown. Provide a 7' deep fluid level in all dipping vats.

The following USDA dipping vat plans are available from the institutions listed on the inside front cover: USDA-5876, Vat for Dipping Cattle, and USDA-5940, Cattle Dipping Vat and Inspection Facility.

Management

Proper vat management prolongs dipping solution life and improves operation efficiency. Management includes providing a clean, debris-free, well agitated, and well aerated dip solution. Solutions allowed to stand too long may give off offensive odors, become septic, and cause infection. Aeration is the most practical form of odor control and solution agitation.

A built-in or portable agitator can be installed. A pipe with 1/16" diameter holes spaced 5" apart can be installed at the bottom of the vat and can be operated while animals are dipped. A portable device can be a pipe handle with a cross pipe at the end that reaches the bottom of the vat. Make the cross piece slightly

shorter than the narrowest part of the vat bottom. Connect the pipe handle to a compressor with a rubber hose long enough to allow the operator to move along the entire vat length. Provide an air compressor capable of delivering 1 cfm/ft of length at 40 psi. Agitate the vat for at least 30 min before dipping any animals. Thoroughly mix the dip solution after each period of nonuse and after each time the vat is replenished.

To maintain effectiveness, periodically clean and recharge the vat, based on the following criteria:
- When 10% of vat volume is displaced by sediment.
- When 2 animals/gal of initial dip solution have been dipped. For example, after 6,000 head have been dipped in a 3,000 gallon vat, clean the vat.
- When the vat has been charged for the maximum number of days recommended for the pesticide used. Check the manufacturer's recommendations.

Keep good records to monitor the dipping vat operation and answer any questions. Make all records readily available to anyone assigned to the dipping operation. Records should include:

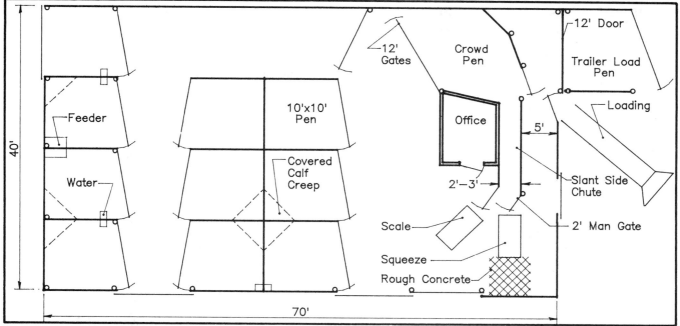

Fig 7-14. Cattle hospital barn.
For calving and small scale cattle feeding. An overhead mow is useful for emergency feed storage.

- Name and location of operation.
- Type and chemical lot number of pesticide used.
- Date vat was charged.
- Number of animals dipped each day and a running total of animals dipped since the vat was charged.
- Solution samples taken and analysis data.
- Name of person supervising dipping.

Safe handling and storage of pesticides is essential. Safety precautions listed on product labels must be understood before use. Provide protective clothing—rubber boots, rubber aprons, unlined rubber gloves—and approved respirators for all persons working with pesticides. Have the telephone number of the nearest poison control center readily available. Provide water and soap to immediately remove spilled or splashed dip materials. Carelessness is the major cause of pesticide accidents. Use caution when handling or storing pesticides.

Dispose of spent dip solution in accordance with label instructions and state and local regulations. Although most dip chemicals are not highly toxic, spreading them around the feedlot may be illegal.

Scale

Scales for legal trade are regularly checked and state approved. A small (about 3′x8′) individual animal scale can weigh cattle for production records. Offset the scale from the working chute or loading chute so it can be bypassed. Large scales (about 9′x20′) are placed outside the cattle handling facilities for weighing other trucks and farm equipment. Separate cattle and commodity scales to avoid congestion, dust, and wear. A combination cattle and commodity scale is at least 10′x60′ with 100,000 lb capacity.

Hospital Pens

Provide 40 to 50 ft²/hd of hospital space for 2%-5% of the finishing and adult cattle. Do not overcrowd sick animals. Provide one hospital area for every 6,000 head. Locate this area close to handling facilities and conditioning lots. Provide separate drainage. Use roughened concrete sloped ¼″/ft or more to a drain for outdoor hospital pens. See Fig 7-14.

More than one hospital/treatment area is recommended for large custom operations. Separate this area to reduce disease transmission. Clean and disinfect before putting new groups of cattle in.

Heat and mechanically ventilate tight, well insulated rooms and intensively used barn areas. Space to drive a veterinarian's truck into the treatment area is desirable.

Treatment Supply Room

Provide a small insulated and heated building or a room near the treatment area to store equipment and supplies used at the squeeze chute and headgate. Provide a refrigerator for veterinary supplies. Lock this room when not in use. Where large numbers of cattle are treated regularly, install a water heater and sink with hot and cold water. Consider an emergency shower head in case of contamination with treatment products. See Figs 2-3, 2-4, 2-5, and 7-5.

Management

Thoroughly clean treating, handling, hospital, and barn areas during the early part of the summer. Use white wash containing cresol (a disinfectant), or equivalent, for washing walls, posts, and other surfaces. Cresol helps control ringworm and lice caused by cattle rubbing against walls. Heavily spread dry lime on floor areas about 30 days before use.

Transport Space Requirements

Table 7-2. Cattle capacity for 92″ wide trailers.
When hauling cattle, check that the legal weight limit of the trailer is not exceeded.

Animal weight, lb		Trailer length, ft						Ft²/hd
		16	18	20	22	44	46	
200	head	35	40	44	48	97	101	3.5
	wt, lb	7,000	8,000	8,800	9,600	19,400	20,200	
300	head	26	29	32	35	70	74	4.8
	wt, lb	7,800	8,700	9,600	10,500	21,000	22,200	
400	head	19	22	24	26	53	55	6.4
	wt, lb	7,600	8,800	9,600	10,400	21,200	22,000	
500	head	16	18	20	22	44	46	7.7
	wt, lb	8,000	9,000	10,000	11,000	22,000	23,000	
600	head	14	16	18	20	40	41	8.5
	wt.lb	8,400	9,600	10,800	12,000	24,000	24,600	
700	head	13	14	16	18	35	37	9.6
	wt, lb	9,100	9,800	11,200	12,600	24,500	25,900	
800	head	11	13	14	15	31	32	11.0
	wt, lb	8,800	10,400	11,200	12,000	24,800	25,600	
900	head	10	12	13	14	29	30	11.8
	wt, lb	9,000	10,800	11,700	12,600	26,100	27,000	
1,000	head	10	11	12	13	26	28	12.8
	wt, lb	10,000	11,000	12,000	13,000	26,000	28,000	
1,200	head	8	9	10	11	22	23	15.3
	wt, lb	9,600	10,800	12,000	13,200	26,400	27,600	
1,400	head	6	7	8	9	18	18	19.2
	wt, lb	8,400	9,800	11,200	12,600	25,200	25,200	

Equipment

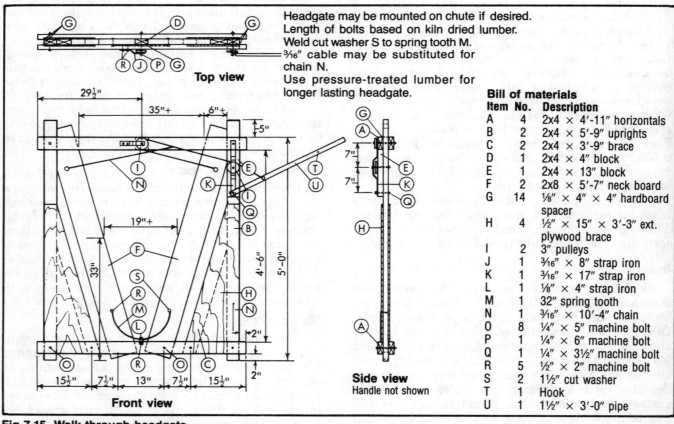

Headgate may be mounted on chute if desired.
Length of bolts based on kiln dried lumber.
Weld cut washer S to spring tooth M.
³⁄₁₆″ cable may be substituted for chain N.
Use pressure-treated lumber for longer lasting headgate.

Top view

Front view

Side view
Handle not shown

Bill of materials

Item No.		Description
A	4	2x4 × 4′-11″ horizontals
B	2	2x4 × 5′-9″ uprights
C	2	2x4 × 3′-9″ brace
D	1	2x4 × 4″ block
E	1	2x4 × 13″ block
F	2	2x8 × 5′-7″ neck board
G	14	1/8″ × 4″ × 4″ hardboard spacer
H	4	½″ × 15″ × 3′-3″ ext. plywood brace
I	2	3″ pulleys
J	1	³⁄₁₆″ × 8″ strap iron
K	1	³⁄₁₆″ × 17″ strap iron
L	1	1/8″ × 4″ strap iron
M	1	32″ spring tooth
N	1	³⁄₁₆″ × 10′-4″ chain
O	8	¼″ × 5″ machine bolt
P	1	¼″ × 6″ machine bolt
Q	1	¼″ × 3½″ machine bolt
R	5	½″ × 2″ machine bolt
S	2	1½″ cut washer
T	1	Hook
U	1	1½″ × 3′-0″ pipe

Fig 7-15. Walk-through headgate.
Use pressure preservative treated lumber for a longer lasting headgate. Mount the headgate on a chute or squeeze.

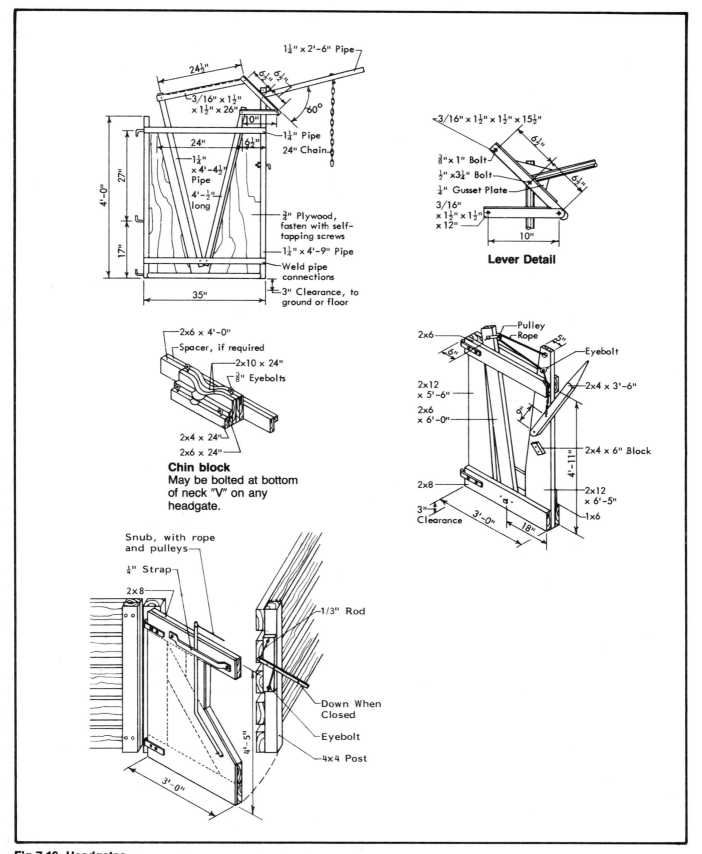

Fig 7-16. Headgates.
Home built headgates are suitable for two-operator use. Commercially built units are available with a self-catching feature for one-operator and intensive use.

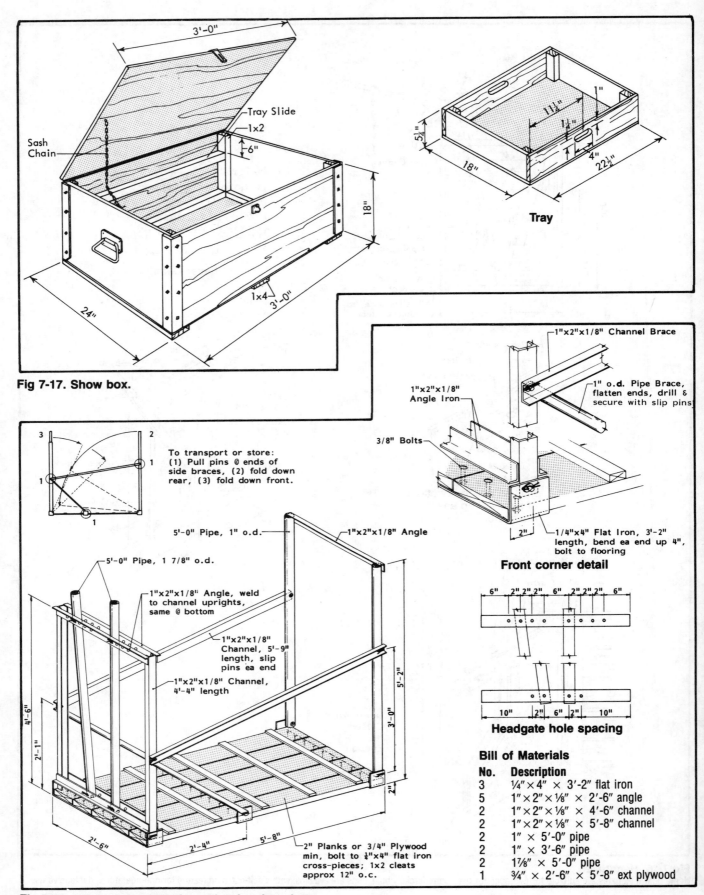

Fig 7-17. Show box.

Tray

To transport or store:
(1) Pull pins @ ends of side braces, (2) fold down rear, (3) fold down front.

5'-0" Pipe, 1" o.d.

5'-0" Pipe, 1 7/8" o.d.

1"x2"x1/8" Angle

1"x2"x1/8" Angle, weld to channel uprights, same @ bottom

1"x2"x1/8" Channel, 5'-9" length, slip pins ea end

1"x2"x1/8" Channel, 4'-4" length

2" Planks or 3/4" Plywood min, bolt to ¼"x4" flat iron cross-pieces; 1x2 cleats approx 12" o.c.

1"x2"x1/8" Channel Brace

1" o.d. Pipe Brace, flatten ends, drill & secure with slip pins

1"x2"x1/8" Angle Iron

3/8" Bolts

1/4"x4" Flat Iron, 3'-2" length, bend ea end up 4", bolt to flooring

Front corner detail

Headgate hole spacing

Bill of Materials

No.	Description
3	¼" × 4" × 3'-2" flat iron
5	1" × 2" × ⅛" × 2'-6" angle
2	1" × 2" × ⅛" × 4'-6" channel
2	1" × 2" × ⅛" × 5'-8" channel
2	1" × 5'-0" pipe
2	1" × 3'-6" pipe
2	1⅞" × 5'-0" pipe
1	¾" × 2'-6" × 5'-8" ext plywood

Fig 7-18. Heavy-duty portable cattle trimming chute.

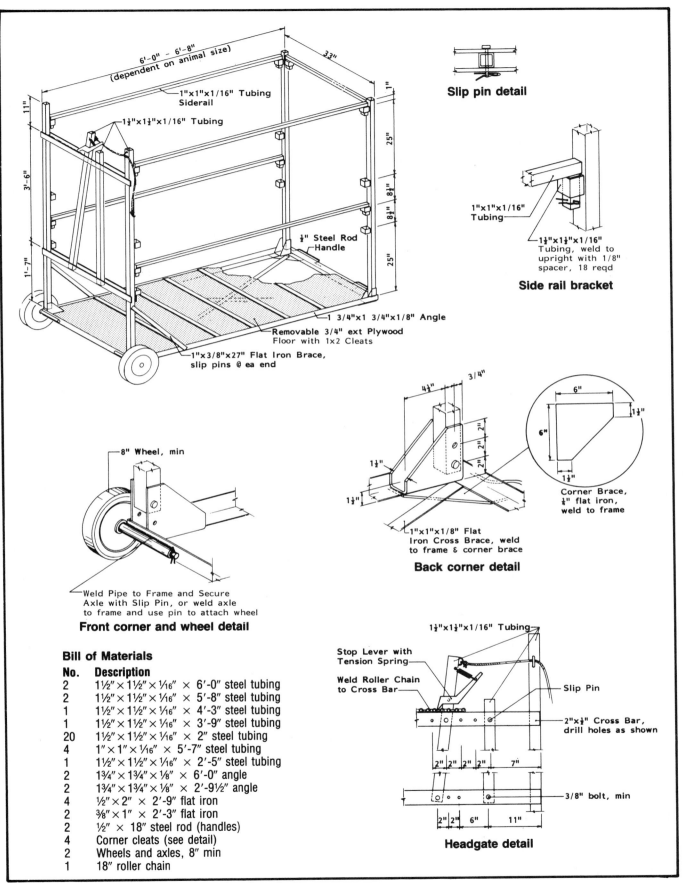

Slip pin detail

Side rail bracket

Front corner and wheel detail

Back corner detail

Headgate detail

Bill of Materials

No.	Description
2	1½″ × 1½″ × 1/16″ × 6′-0″ steel tubing
2	1½″ × 1½″ × 1/16″ × 5′-8″ steel tubing
1	1½″ × 1½″ × 1/16″ × 4′-3″ steel tubing
1	1½″ × 1½″ × 1/16″ × 3′-9″ steel tubing
20	1½″ × 1½″ × 1/16″ × 2″ steel tubing
4	1″ × 1″ × 1/16″ × 5′-7″ steel tubing
1	1½″ × 1½″ × 1/16″ × 2′-5″ steel tubing
2	1¾″ × 1¾″ × 1/8″ × 6′-0″ angle
2	1¾″ × 1¾″ × 1/8″ × 2′-9½″ angle
4	½″ × 2″ × 2′-9″ flat iron
2	3/8″ × 1″ × 2′-3″ flat iron
2	½″ × 18″ steel rod (handles)
4	Corner cleats (see detail)
2	Wheels and axles, 8″ min
1	18″ roller chain

Fig 7-19. Lightweight portable cattle trimming chute.

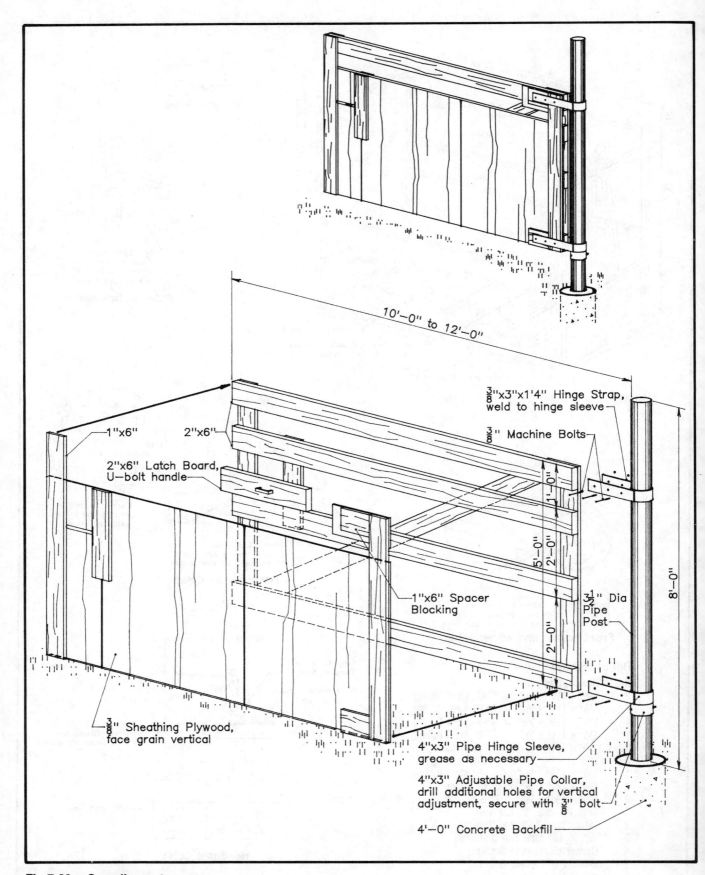

1"x6"

2"x6"

2"x6" Latch Board,
U—bolt handle

10'—0" to 12'—0"

$\frac{3}{8}$"x3"x1'4" Hinge Strap,
weld to hinge sleeve

$\frac{3}{8}$" Machine Bolts

1"x6" Spacer
Blocking

$3\frac{1}{2}$" Dia
Pipe
Post

1'—0"

5'—0"

2'—0"

2'—0"

8'—0"

$\frac{3}{8}$" Sheathing Plywood,
face grain vertical

4"x3" Pipe Hinge Sleeve,
grease as necessary

4"x3" Adjustable Pipe Collar,
drill additional holes for vertical
adjustment, secure with $\frac{3}{8}$" bolt

4'—0" Concrete Backfill

Fig 7-20. Crowding gate.

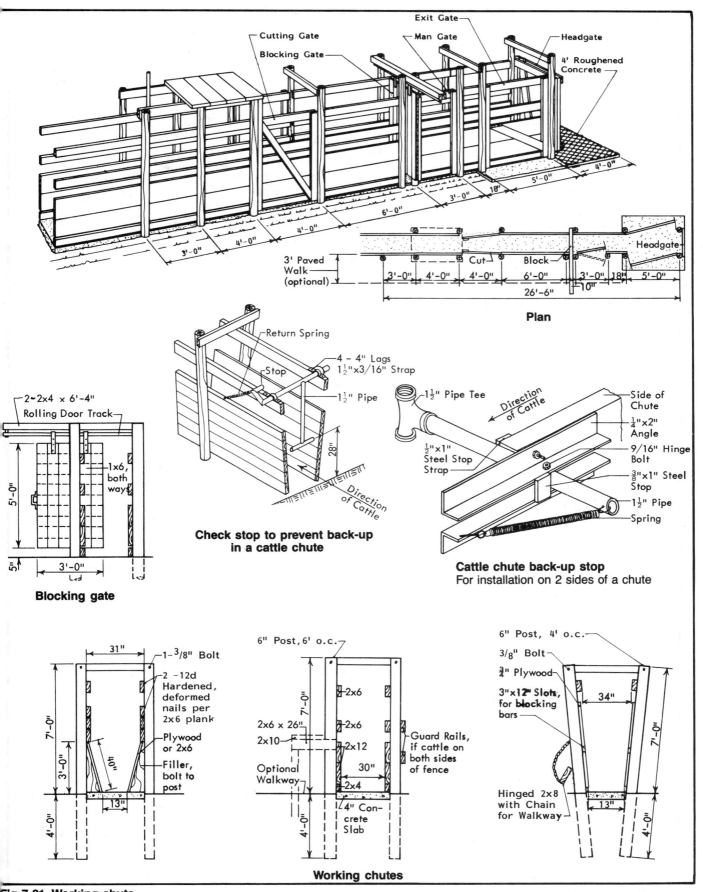

Fig 7-21. Working chute.
A platform mounted over the cutting gates permits the gate operator to watch cattle approaching the gates. It also provides for other observers out of the way of cattle handlers.

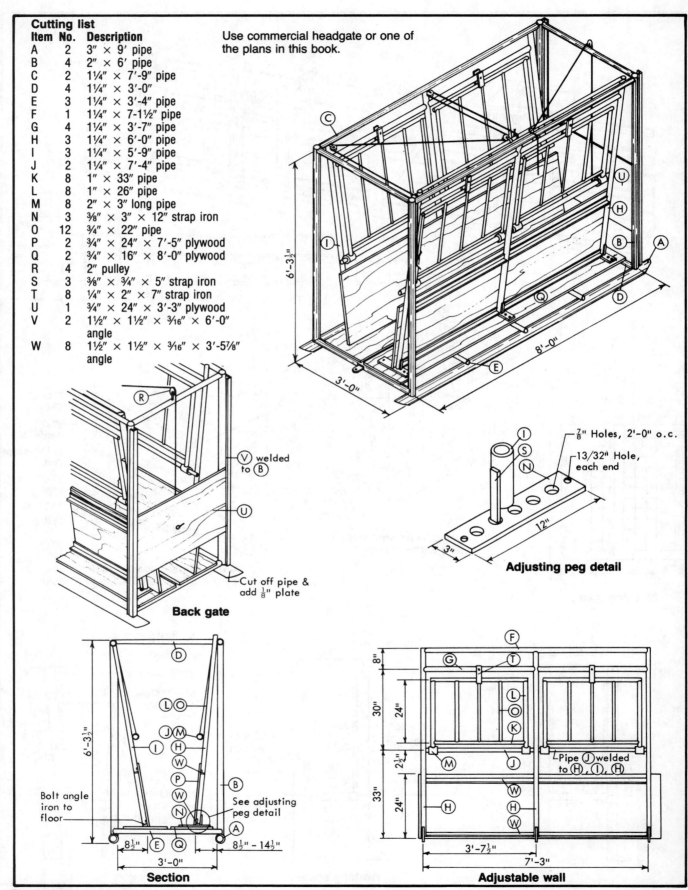

Cutting list

Item	No.	Description
A	2	3″ × 9′ pipe
B	4	2″ × 6′ pipe
C	2	1¼″ × 7′-9″ pipe
D	4	1¼″ × 3′-0″
E	3	1¼″ × 3′-4″ pipe
F	1	1¼″ × 7-1½″ pipe
G	4	1¼″ × 3′-7″ pipe
H	3	1¼″ × 6′-0″ pipe
I	3	1¼″ × 5′-9″ pipe
J	2	1¼″ × 7′-4″ pipe
K	8	1″ × 33″ pipe
L	8	1″ × 26″ pipe
M	8	2″ × 3″ long pipe
N	3	⅜″ × 3″ × 12″ strap iron
O	12	¾″ × 22″ pipe
P	2	¾″ × 24″ × 7′-5″ plywood
Q	2	¾″ × 16″ × 8′-0″ plywood
R	4	2″ pulley
S	3	⅜″ × ¾″ × 5″ strap iron
T	8	¼″ × 2″ × 7″ strap iron
U	1	¾″ × 24″ × 3′-3″ plywood
V	2	1½″ × 1½″ × 3⁄16″ × 6′-0″ angle
W	8	1½″ × 1½″ × 3⁄16″ × 3′-5⅞″ angle

Use commercial headgate or one of the plans in this book.

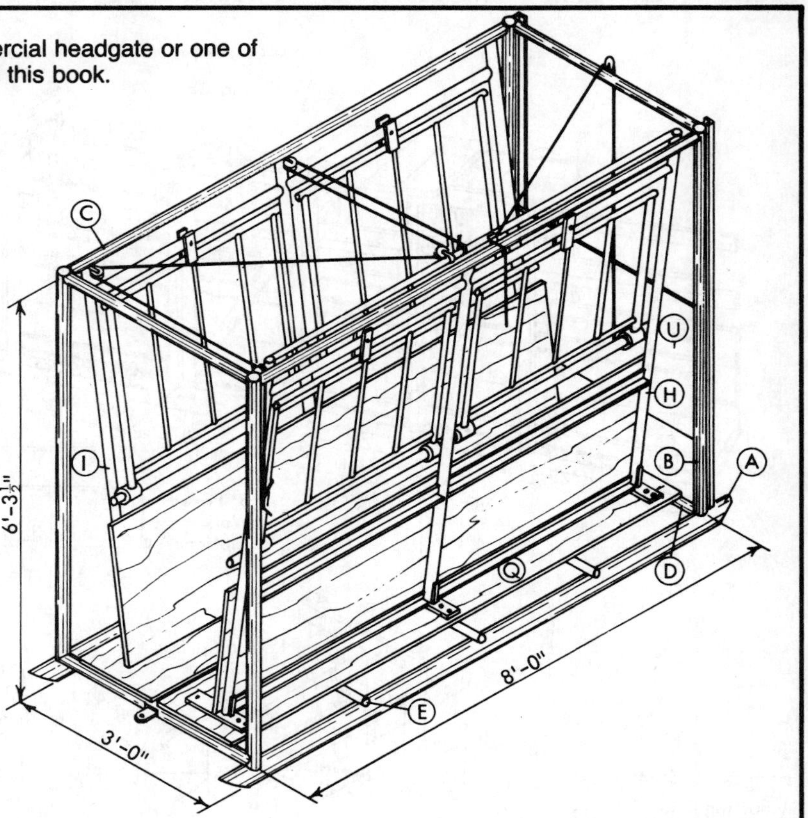

C
U
H
I
B
A
Q
D
6′-3½″
8′-0″
3′-0″
E

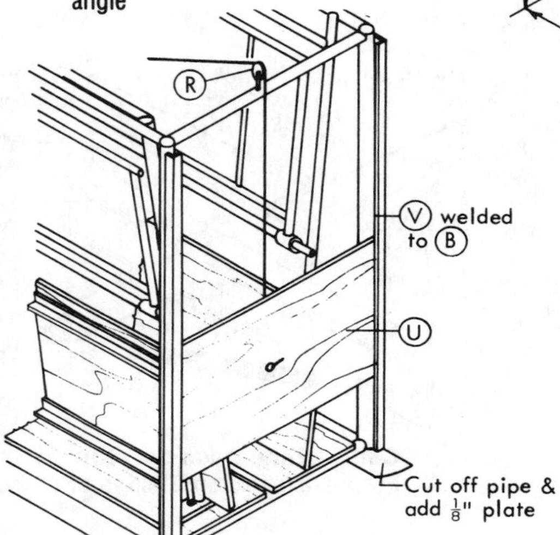

R
V welded to B
U
Cut off pipe & add ⅛″ plate

Back gate

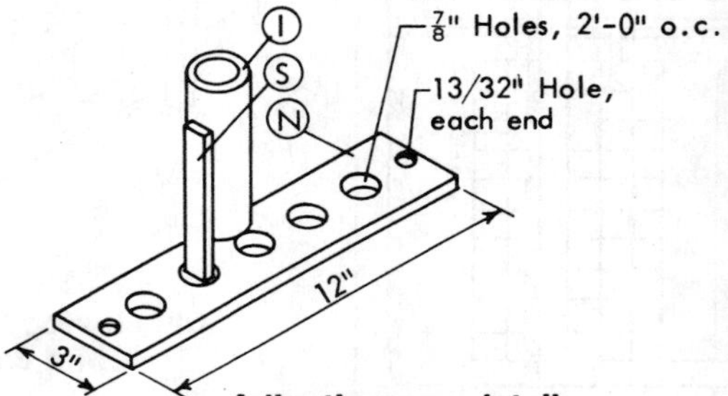

I
S
N
⅞″ Holes, 2′-0″ o.c.
13/32″ Hole, each end
12″
3″

Adjusting peg detail

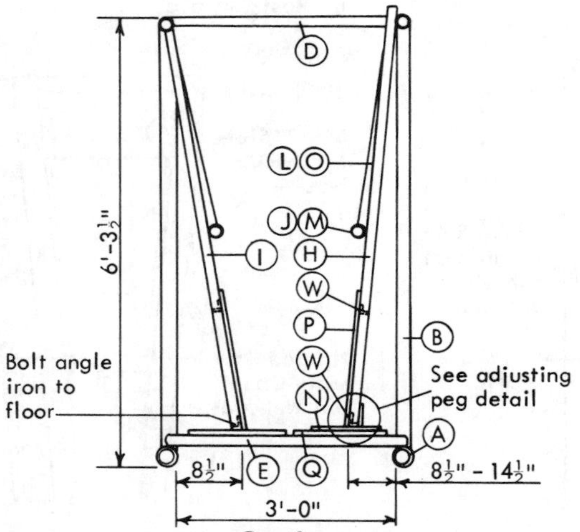

D
L O
J M
I
H
W
P
W
N
B
A
6′-3½″
Bolt angle iron to floor
See adjusting peg detail
8½″
E
Q
8½″ - 14½″
3′-0″

Section

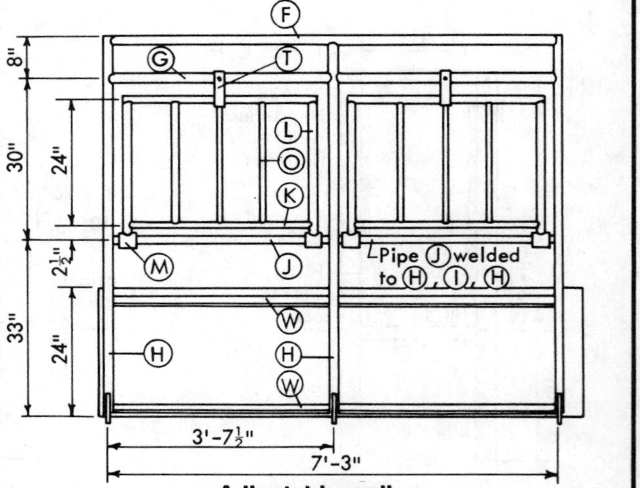

F
G
T
8″
30″
24″
L
O
K
2½″
M
J
Pipe J welded to H, I, H
33″
24″
H
W
H
W
3′-7½″
7′-3″

Adjustable wall

Fig 7-22. Cattle squeeze.
Provide a headgate to catch and restrain the animal. Use a commercially available headgate or a plan from this book.

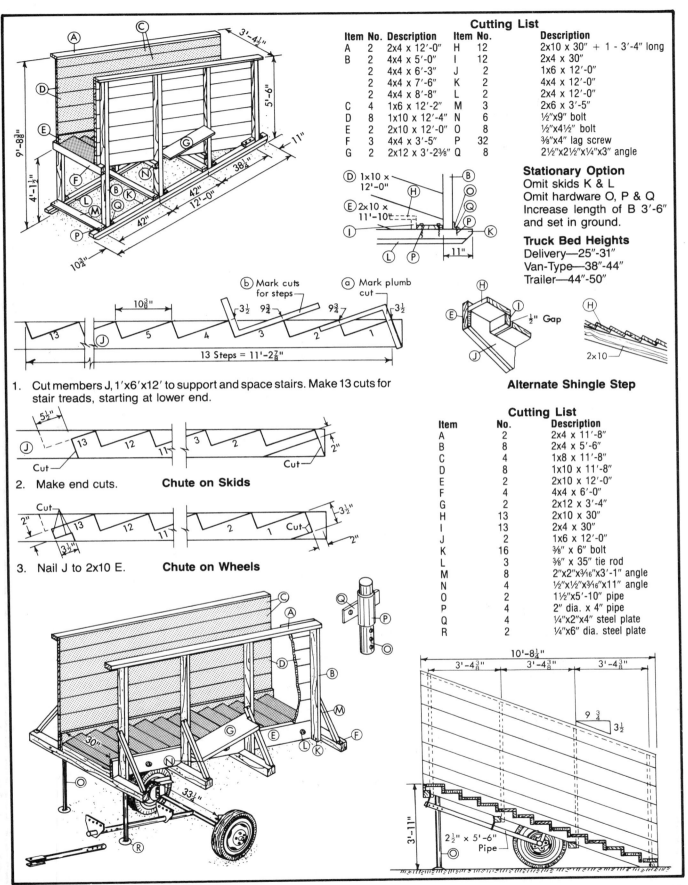

Cutting List

Item	No.	Description	Item	No.	Description
A	2	2x4 x 12'-0"	H	12	2x10 x 30" + 1 - 3'-4" long
B	2	4x4 x 5'-0"	I	12	2x4 x 30"
	2	4x4 x 6'-3"	J	2	1x6 x 12'-0"
	2	4x4 x 7'-6"	K	2	4x4 x 12'-0"
	2	4x4 x 8'-8"	L	2	2x4 x 12'-0"
C	4	1x6 x 12'-2"	M	3	2x6 x 3'-5"
D	8	1x10 x 12'-4"	N	6	½"x9" bolt
E	2	2x10 x 12'-0"	O	8	½"x4½" bolt
F	3	4x4 x 3'-5"	P	32	⅜"x4" lag screw
G	2	2x12 x 3'-2⅜"	Q	8	2½"x2½"x¼"x3" angle

Stationary Option

Omit skids K & L
Omit hardware O, P & Q
Increase length of B 3'-6"
and set in ground.

Truck Bed Heights

Delivery—25"-31"
Van-Type—38"-44"
Trailer—44"-50"

1. Cut members J, 1'x6'x12' to support and space stairs. Make 13 cuts for stair treads, starting at lower end.

2. Make end cuts. **Chute on Skids**

3. Nail J to 2x10 E. **Chute on Wheels**

Alternate Shingle Step

Cutting List

Item	No.	Description
A	2	2x4 x 11'-8"
B	8	2x4 x 5'-6"
C	4	1x8 x 11'-8"
D	8	1x10 x 11'-8"
E	2	2x10 x 12'-0"
F	4	4x4 x 6'-0"
G	2	2x12 x 3'-4"
H	13	2x10 x 30"
I	13	2x4 x 30"
J	2	1x6 x 12'-0"
K	16	⅜" x 6" bolt
L	3	⅜" x 35" tie rod
M	8	2"x2"x³⁄₁₆"x3'-1" angle
N	4	½"x½"x³⁄₁₆"x11" angle
O	2	1½"x5'-10" pipe
P	4	2" dia. x 4" pipe
Q	4	¼"x2"x4" steel plate
R	2	¼"x6" dia. steel plate

Fig 7-23. Portable loading chutes.

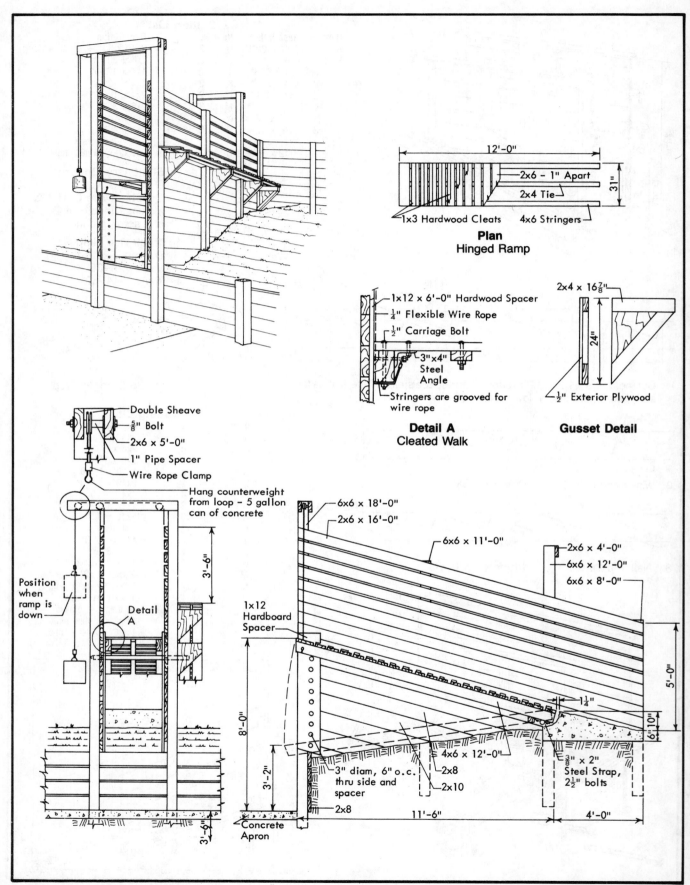

Fig 7-24. Variable height loading chute.

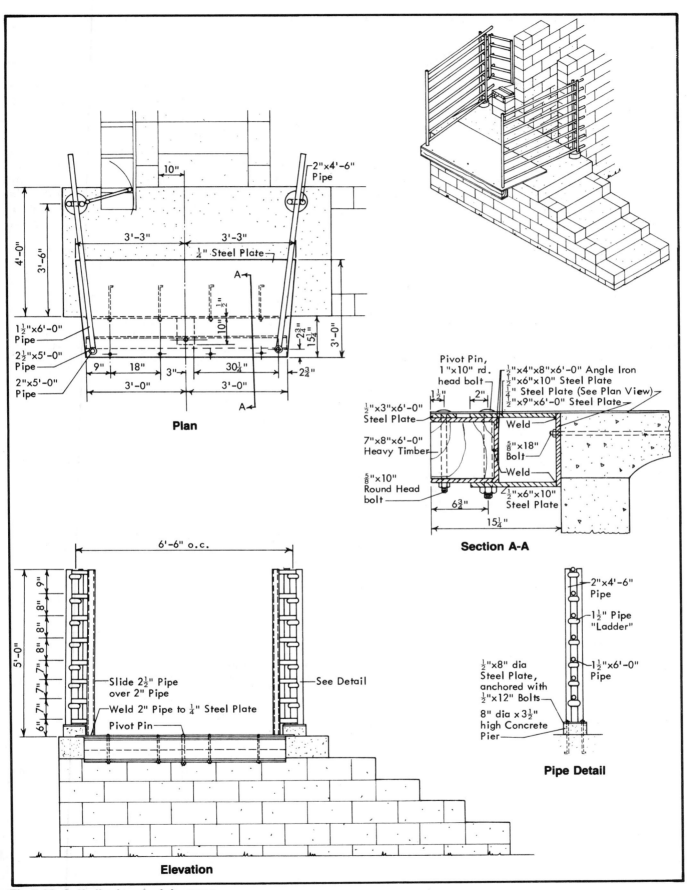

Plan

10"

2"x4'-6" Pipe

3'-3" 3'-3"

¼" Steel Plate

4'-0"

3'-6"

A

½"

10"

2¾"
15¼"

3'-0"

1½"x6'-0" Pipe

2½"x5'-0" Pipe

2"x5'-0" Pipe

9" 18" 3" 30¼" 2¾"

3'-0" 3'-0"

A

Section A-A

Pivot Pin, 1"x10" rd. head bolt

½"x4"x8"x6'-0" Angle Iron
½"x6"x10" Steel Plate
¼" Steel Plate (See Plan View)
½"x9"x6'-0" Steel Plate

1½"

2"

½"x3"x6'-0" Steel Plate

Weld

7"x8"x6'-0" Heavy Timber

⅝"x18" Bolt

Weld

⅝"x10" Round Head bolt

½"x6"x10" Steel Plate

6¾"

15¼"

Elevation

6'-6" o.c.

5'-0"

9"

8"

8"

8"

7"

7"

7"

6"

See Detail

Slide 2½" Pipe over 2" Pipe

Weld 2" Pipe to ¼" Steel Plate

Pivot Pin

Pipe Detail

2"x4'-6" Pipe

1½" Pipe "Ladder"

1½"x6'-0" Pipe

½"x8" dia Steel Plate, anchored with ½"x12" Bolts

8" dia x 3½" high Concrete Pier

Fig 7-25. Self aligning dock bumper.
Platform matches up the chute floor with the truck or trailer floor preventing openings that frighten cattle. See Table 7-1 for recommended height.

8. FEED STORAGE, PROCESSING, AND HANDLING

Factors to consider for feed storage, processing, and handling:
- Number of animals fed.
- Ration.
- Type and size of grain and roughage storage.
- Equipment for unloading, metering, mixing, and processing feed.
- Conveyors and elevators to collect and transport feed.
- Location and condition of existing storages.
- Electrical requirements—certain equipment requires 240 volt or 3-phase service.
- Future expansion.

Feeding Methods

Feed type, particularly hay, affects the type of feeding equipment used. Loose hay, rectangular bales, and big round bales can be fed in self feeders or spread on clean ground. Dry-chopped hay, haylage, silage, and grain require a bunk. Ration determines required feed processing equipment, feed transport equipment, access roads, storages, and bunks.

Three feeding methods are:
- Fenceline feeding.
- Mechanical bunk feeding.
- Self feeding.

Fenceline bunks are along the side of a lot. Fill bunks with side unloading wagons or truck mounted feeder boxes from an access road outside the lot. Vehicles and cattle are separated. Provide all-weather, year-round access. Fenceline bunks require twice as much bunk length (but not twice the cost) as bunks that feed from both sides.

Mechanical bunks usually allow cattle to eat from both sides and can be lot dividers. Mechanical bunks are common with continuous feed processing systems and small (about 300 head) operations. Limit bunk length to 200'. Conveyors and mechanical bunks reduce feeding time and labor but are more expensive than fenceline bunks. Cover mechanical bunks to protect equipment and feed. Consider operating and maintenance costs when selecting a feeding system. Provide a backup system in case of electrical or equipment failure.

Self feeding systems include haystacks, silage from horizontal silos or plastic bags, and grain and mixed rations in bunks or self feeders. Feeders can be filled with portable mixer wagons, truck mounted feeder boxes, or overhead augers from outside the lot.

Grain Storage

Grain can be stored at harvest or bought when needed. The biggest bin is not always the least costly per bushel of storage. To allow storing different grains, using fresher grain, and more flexibility, the biggest bin should not exceed half of the total volume of the operation. For freshly harvested grain, equip storage bins with aeration to cool grain. For more information on grain aeration, obtain *Managing Dry Grain in Storage*, AED-20, by Midwest Plan Service. Smaller facilities are needed if grain and other feed ingredients are purchased as required. See MWPS-13, *Grain Drying, Handling, and Storage Handbook*, for more information.

Hopper-bottomed bins permit gravity unloading of stored materials before and after processing, Fig 8-1. Side draw-off hoppers are for materials that tend to bridge, such as concentrates and ground feeds. Steep sideslopes allow these materials to flow more freely. Off-center unloading can cause denting at the sidewall, so limit bulk feed tank capacity to 20 tons.

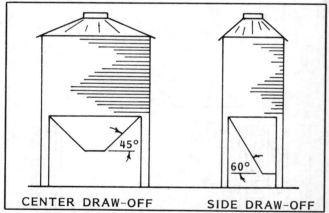

CENTER DRAW-OFF   SIDE DRAW-OFF

Fig 8-1. Hopper-bottomed bins.
Center draw-off bins work best for whole grain and free flowing materials. The steeper sloped side draw-off bins are for ground feed. Agitators or vibrators are available to assist gravity flow.

Table 8-1. Feeder alternatives for cattle.

Type of feeder	Cost	Loose hay	Ground, dry hay	Bales Rectan-gular	Bales Big round	Silage, haylage	Grain or pellets	Feed handling system recommended
Portable bunk	Medium	Some	Yes	Some	—	Yes	Yes	Hand feed
Fenceline bunk	High	No	Yes	Some	Grind	Yes	Yes	Front-end loader, unloader wagon
Self feeder	Medium	Yes	Yes	Yes	Yes	Some	Some	Front-end loader, unloader wagon
Mechanical bunk	Highest	No	Yes	No	No	Yes	Yes	Upright silos with conveyors

Hay Storage

Store hay near loading or feeding areas. Some hay stacks can be used as windbreaks. The amount of spoilage—as much as 50%—depends on the stack or bale quality, storage method, and local rainfall. Indoor storage is recommended in areas with more than 20″ of annual rainfall. For outdoor storage, provide a gentle slope and at least a 1′ space between stacks or bales to promote drainage and air movement. Divert surface runoff away from storage areas.

Hay Feeding

With self feeding, waste can be as much as 50%. Trampling creates mud and usually kills vegetation when stacks are self fed in the hay field. Reduce hay waste to 5%-10% by controlling animal access to hay.

- Fence storages on 3 sides; use electric fence along the fourth side to control access and hay consumption, Fig 8-3. Allow 4″-6″ of space/head when hay is always available and 18″-24″/head for limited feeding.
- Place large bales and stacks in lines 15′ apart and 8′-10′ from a permanent fence. Move temporary fencing to expose one bale or stack at a time.
- Self feed whole bales or stacks in 3-sided fence-line feeders or a push-up feeding fence, Fig 8-4.
- Self feed whole bales or stacks in circular feeders, feeder panels, or portable feeders, Fig 8-6.
- Limit feed with a bale unroller or a feeder attachment on a trailed stack mover. Drop the hay on the ground or into bunks, Figs 8-5 and 8-6.

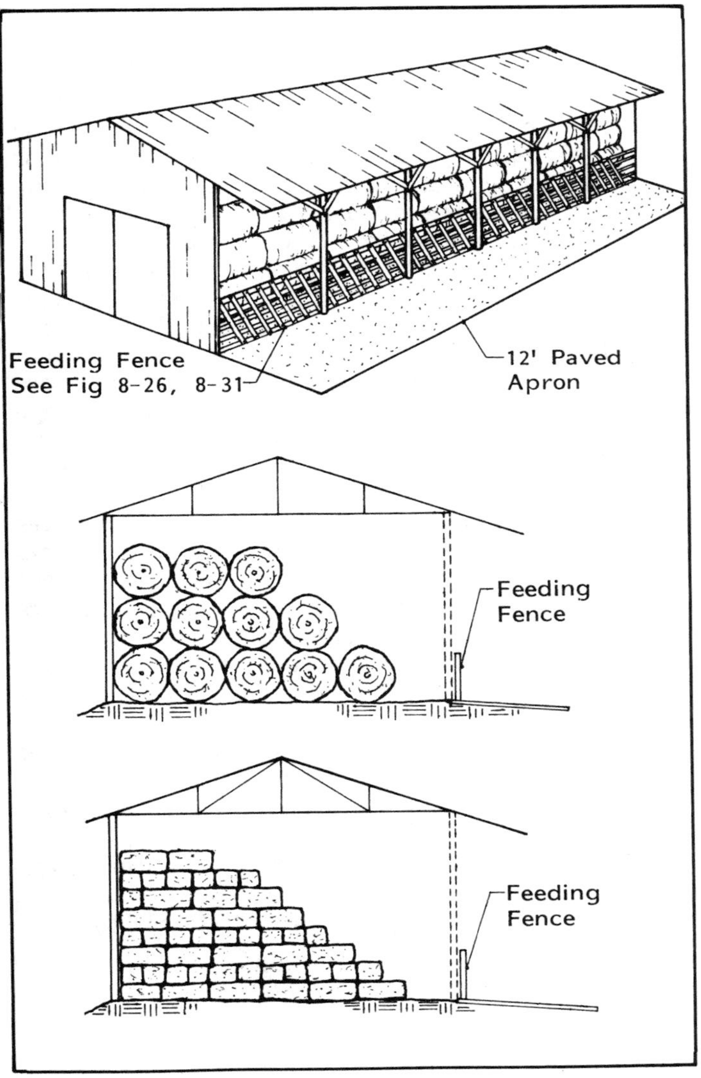

Feeding Fence
See Fig 8-26, 8-31

12′ Paved Apron

Feeding Fence

Feeding Fence

Fig 8-2. Hay feeding from roofed storage.
Do not stack bales against sidewalls unless building is designed for the loads imposed by storage. Indoor hay storage helps preserve quality and reduce dry matter losses. The following Midwest Plan Service hay barn plans are available: mwps-73110, 24′ Wide Hay Barn; mwps-73111, 36′ Wide Hay Barn; mwps-73112, 48′ Wide Hay Barn; mwps-73113, 32′ and 48′ Wide Hay Sheds.

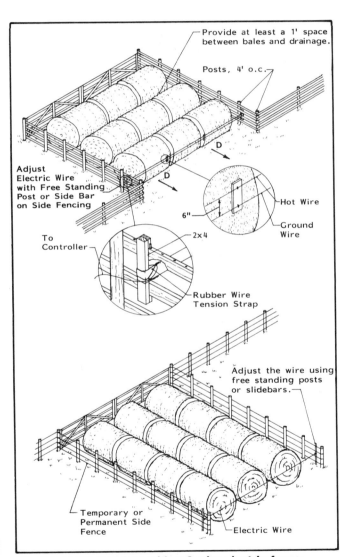

Provide at least a 1′ space between bales and drainage.

Posts, 4′ o.c.

D

Adjust Electric Wire with Free Standing Post or Side Bar on Side Fencing

Hot Wire

6″

Ground Wire

To Controller

2x4

Rubber Wire Tension Strap

Adjust the wire using free standing posts or slidebars.

Temporary or Permanent Side Fence

Electric Wire

Fig 8-3. Rationing hay with a 2-wire electric fence.
Animals feed under the hot wire, which is moved as required to control consumption. Slope away from the electric feeding fence to promote drainage. Provide drainage between and under bales if stored at the site for several months.

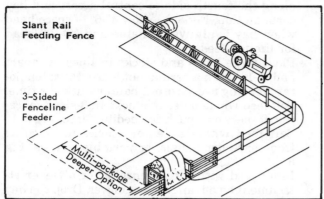

Fig 8-4. Fenceline feeding of big bales and stacks.
Push the hay closer, as necessary. Clean manure away weekly. Frozen buildup can cause problems.

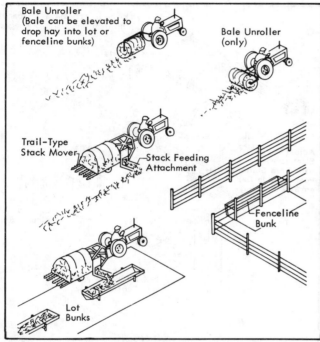

Fig 8-5. Big bale and stack limit feeding methods.
Feed on clean ground or in fenceline bunks.

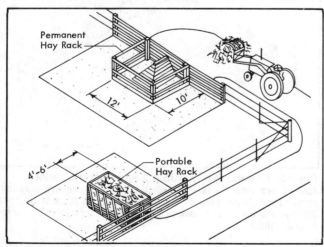

Fig 8-6. Limit feeding roughage.
Fill feeders with a front-end loader or by hand for a small operation.

Silage Storage

Horizontal silos may be most economical for very small producers, upright silos for the intermediate operation, and big horizontals for the large operator. Operator preference, ration, and feeding circumstances affect selection.

Upright silos with unloaders offer protection from the weather and possible complete mechanization of silage handling. They can be conventional silos with top unloaders or air limiting silos with bottom unloaders. Conventional silos cost less per ton of dry matter stored, but the top unloader must be raised during filling, which interrupts feeding. Conventional silos require tight walls and sealed doors. A roof or plastic seal over the silage reduces storage losses. The initial cost of air limiting silos is high, but they offer more flexibility because refilling does not interfere with the unloader or feeding.

Air limiting silos usually have lower storage losses but higher field losses because of low harvesting moisture content. Spoilage losses in conventional silos can be high when only a thin layer is removed each day, silage is not fed for several days, or the surface is left uncovered. Once a conventional silo is opened, feed out at least 4"/day in warm weather.

Horizontal silos include bunkers, trenches, and free-standing stacks. Use bunkers and stacks on level sites and trenches on sloping sites. Horizontal silos cost less per stored ton than upright silos but have higher storage losses unless carefully managed. Sidewall heights are usually 8'-14' and widths are 30'-60'. See Tables 8-3 and 8-4 to estimate storage capacity. See AED-15, *Tilt-Up Concrete Horizontal Silo Construction,* for design and construction guidelines.

Proper management of horizontal silos reduces storage losses. Wind, rain, snow, birds, and rodents are problems. Chop corn and hay at ¼" theoretical cut with 65%-70% moisture content, wet basis. Minimize surface exposure during filling; fill one end nearly full quickly. Pack horizontal silos continuously with a wheel-type tractor while filling and periodically for 2 or 3 days after filling. Shape the silage surface to shed water. Cover with heavy black plastic and weight down with old tires laid side by side, Fig 8-7. Black plastic resists ultraviolet deterioration. Deep silos have less surface spoilage. Narrow silos have less face spoilage because a thicker layer is removed each day.

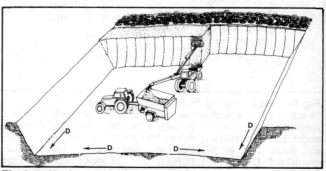

Fig 8-7. Horizontal silo unloader.
Unload bunker or trench silos with minimal losses.

Horizontal silos adapt readily to feed wagon distribution and to high speed unloading with tractor loaders or horizontal silo unloaders. Unload 4" daily and maintain a hard surface floor sloped for drainage to control spoilage.

Horizontal plastic bags can be used for shortor long-term storage at several locations or tenant farms, or for emergency storage, Fig 8-8. Plastic bag silos can be in protected areas for winter feeding and in pastures or lots for spring feeding.

A portable filler/bagging machine fills 8' diameter by about 150' long plastic bags. Bags can be partially filled, closed, and later completely filled. Keep bags tightly closed because billowing plastic pumps air over the silage, increasing losses. Maximum moisture content of 50% reduces freezing and spoilage problems. Moisture contents of 60%-70% promote more fermentation but can freeze solid.

Bags are not reusable and must be protected from punctures. Common causes of spoilage are tears by rodents or animals, splits at the seam, punctures from objects on the ground, and weather damage.

A firm, well drained site is essential for year-round access. Orient bags north-south to promote melting and drying on each side. Provide a 5% slope away from the bags for good drainage. A fenced and paved storage area is recommended for self feeding, loading, and unloading in poor weather. Unload from either end of the bag. A specially designed feeding fence on each end self feeds about 30 head.

Stack silage only for emergency or temporary storage. Corn silage stacks are fairly successful with material at 70% moisture. Make stacks as small and as deep as is safe and practical to improve quality. Protect silage with 6 mil black plastic.

Losses for various silos are in Fig 8-9. Measuring storage losses is difficult and values shown are estimates. In some cases, little dry matter loss but substantial feeding quality loss can occur.

Determining Silo Size

Silo capacity depends on height and diameter. Crop moisture is also important, but silo capacity on a dry matter basis is consistent between 50%-70% moisture content.

Silo height

Determine the minimum silo height based on required silage removal to prevent spoilage, feeding period, and freeboard. See Table 8-10 for suggested removal rates. Provide 1' of freeboard for settling in silos up to 30' high and an additional foot for each 10' above 30'. If the silo unloader is stored in the silo, add 10' of freeboard to the height. Calculate the minimum silo height from:

Eq 8-1.

$$H = SR \times FP \div 12 \text{ in/ft} + FB$$

H = minimum silo height, ft
SR = silage removal, in/day, Table 8-10
FP = feeding period, days
FB = freeboard, ft

Silo diameter

Minimum silo height and required storage capacity determine desired silo diameter. Convert silage, wet basis, to dry matter content to determine storage capacity. Calculate the tons of dry matter.

Eq 8-2.

$$DM = S \times FP \div (2,000 \text{ lb/ton} \times DMF)$$

DM = tons of dry matter
S = silage fed, lb/day
FP = feeding period, days
DMF = dry matter factor, Table 8-11

The dry matter factor converts moist silage to dry matter content. Estimate the "as harvested" moisture and select the dry matter factor from Table 8-11. Corn

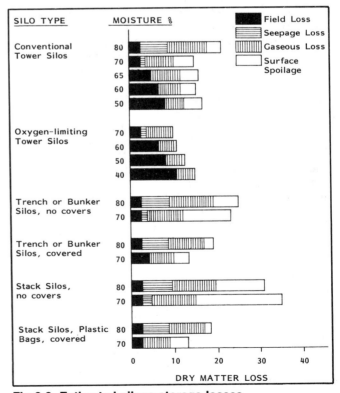

Fig 8-9. Estimated silage storage losses.
Adapted from *Forages*, 3rd edition, Iowa State University Press, Ames, Iowa, 1973.

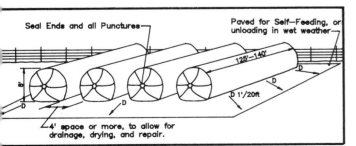

Fig 8-8. Horizontal plastic bag.

silage is usually stored at about 65% and haylage at about 50%.

Select the maximum diameter from Table 8-7 using the minimum height and tons of dry matter. Silos with greater height and smaller diameter that meet the needed storage can also be used. If silage is stored at 65% moisture, determine silo size from Table 8-6 without converting to tons of dry matter. Two or more silos increase handling and management flexibility.

Silage Feeding

Mechanical conveyor feeding best serves feedlots near upright silos with unloaders. Match unloader and conveyor capacities. Limit conveyor lengths to about 100' for augers and 200' for belts. Accurately measuring daily feed with continuous conveyor feeding requires flow through weighers in the system.

Conventional upright silos can be combined with one or more air limiting units. Use conventional tower silos for bulk storage, such as corn silage or corn and cob meal silage. Use an air-limiting unit to ensile roughage and grains added during the summer and early fall, and then fill the silo with wet grain or silage for winter storage. It can also be refilled from other silage storages to use its mechanical unloading capacity.

The equipment to supply a mechanical conveyor bunk can also fill feed wagons or trucks, which are more flexible than mechanical bunks. They can serve different locations, and enterprises (cattle on pasture, feedlots), get feed from off- and on-farm sources, and permit feedlot locations to fit topography, use existing facilities, etc. Vehicle feed distribution adapts to changing cattle numbers and permits knowing how much feed is going to each lot, either by weighing or estimating the volume of each load. Provide all-weather access for feed wagons and trucks.

Silo unloaders are usually too slow for well designed wagon feeding systems. The 10 to 20 min wait to fill a load may be acceptable for one load per feeding, but probably not for two or more loads. A surge bin or accumulator box reduces the delay. The box, mounted on a scale or with a high capacity transfer conveyor, Fig 8-11, accumulates a load while the first load is being delivered. Size the accumulator to the feed wagon box; select one with fast delivery of non-free flowing materials. A surge bin is expensive and usually limited to larger operations.

Feed Processing

Unloading

Unloading grain from storage for feed processing is a low capacity operation. Slow speed conveyors use less power and do not wear as quickly. If grain storage is also used for cash grain, high volume unloading equipment is needed. Provide a surge bin at the mill to eliminate frequent starting and stopping of unloading equipment.

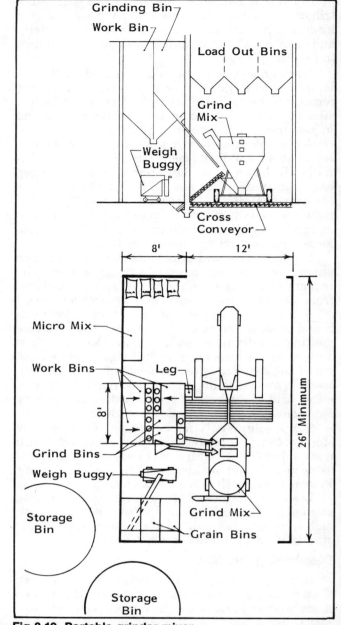

Fig 8-10. Portable grinder-mixer.
Provide an exhaust system for the tractor to the outside to prevent fire.

Processing

Consider a portable grinder-mixer for small lots, a scale-mixer trailer or wagon for 300-1,500 head, and a scale-mixer truck to feed 1,500-5,000 head. One truck is needed for each 5,000 head on feed.

Portable grinder-mixer

Portable grinder-mixers can serve several locations and process feed in batches. One unit grinds, mixes, and delivers at about one batch/hr. They can be equipped to grind and handle hay. Because the mixer can be moved from farm to farm, it is especially suitable for tenant farmers.

Portable grinder-mixers are usually trailer-mounted and pto-powered. Mill selection depends on the type of livestock and kind of feed processed. Large models hold about 120 bu (3 ton). Material is conveyed into a grinder either with augers or chain-slat conveyors. Chain-slat conveyors are used mainly for handling roughage.

Higher labor and operating costs are the major disadvantages of portable grinder-mixers. Both a tractor and operator are required and all ingredients must be weighed into the unit separately. About 40 hp is required for most portable grinder-mixers, but some use 60 hp or more.

Select a portable grinder-mixer with features that save time and labor, such as large hoppers and high grinding capacity. Electronic scales are convenient, improve accuracy, and are an essential part of the grinder-mixer. Other available options are a liquid-molasses attachment, extra inspection windows, unloading auger extensions, and a hopper for adding protein supplement or premixed rations. Evaluate safety equipment and maintenance requirements carefully before deciding. Do not put protein supplement or premixed rations through the grinder.

Electric blender-grinder

Electric blender-grinders are reliable and accurate if kept calibrated. They have low labor, capacity, and power requirements, and low operating costs. Different size units are available that measure, grind, and mix ingredients simultaneously and continuously but do not handle roughages. Grinding and feed distribution can easily be automated.

Store grain and other ingredients close to the grinder. A separate feed delivery system is required. Consider the required electrical service and horsepower before installation. Required horsepower ranges from 2 hp to about 25 hp.

Electric batch mill

With a batch mill, each ingredient is weighed and ground separately giving good control over feed ration composition. But batch milling is expensive and time consuming and requires careful planning to avoid bottlenecks. Although you can use any type of grinder, you may have to adjust a roller mill before processing each ingredient. Electrical service requirements are likely to be greater than for a blender-grinder. For more than one ingredient, a blender is also needed.

Hay mills and tub grinders

Hay and straw can be handled with a forage harvester, stacking wagon, or in large or small bales. Although more wasteful, self feeding roughage eliminates processing. However, processing roughage and mixing it with the grain allows more even consumption.

Hay mills shred and screen solid bales of hay or straw. They are expensive, have a capacity of 6 to 20 tons/hr through a 1″ screen, and require from 25 to 150 hp.

Tub grinders are usually pto-powered with a 75 to 100 hp tractor. A large tub rotates slowly and feeds a

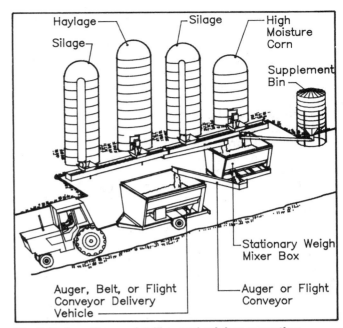

Fig 8-11. Loading, weighing, and mixing operation.
Provide a building over processing area to protect equipment. Building can also be used for feeder wagon storage.

constant supply of roughage to a hammermill in the floor of the grinder. After the roughage is screened, it can be piled on the ground or loaded in a wagon for distribution. A tub grinder is usually fed with a front-end loader and handles about 10 to 12 tons/hr. Processing rate depends on roughage moisture content, screen size, and power available.

Feed Center

A feed center provides for receiving, drying, storing, unloading, elevating and conveying, and processing. Batch feed centers have overhead bins, stationary grinding mills, stationary mixers, and conveyors. Continuous-flow centers have overhead bins feeding an automatic electric mill and conveyors that process and blend ingredients as they flow through the system.

Locate storages for convenient filling, unloading, handling, and mixing. Storage and equipment selection and storage location depend on the feeding system. Conveying wet or dry grain is usually easier than conveying silage and generally requires smaller, less expensive equipment. Locate silos for efficient silage handling and grain storages to allow for post-storage processing and handling.

Locate conveying and processing equipment and the stationary scale mixer in a feed center separated from the barn to reduce interference with natural ventilation. A separate feed center allows for loading a mixed batch into a feed wagon for delivery to animals at other locations.

Store forages and high moisture grains in upright silos and dry grains and concentrates in bins with auger unloaders. Store minerals and vitamin premixes in sacks and weigh them separately ahead of the scale mixer for blending. A conveyor delivers the mixed batch to a mechanical bunk feeder in the barn or lot.

With a mobile mixer, forages and high moisture grains can be stored in upright or horizontal silos. Dry grains and concentrates can be in flat bulk storages. Provide several bays, each with a door and capacity to handle a truckload of grain, premix, etc. Use the same equipment to unload horizontal silos and flat storage bays into the mixer.

Plan for future expansion. Provide space for additional storage and expansion of the feed center building. Layout patterns for silage feed centers and distribution systems are shown in Figs 8-13 to 8-16.

Safety

Augers are one of the most dangerous items on the farm. Cover all auger intakes with a grate designed to keep hands, feet, and clothing from contacting the auger. Make the grate strong enough to support a person. Follow OSHA standards in equipment selection and installation.

Anyone in a grain bin when the unloading auger is running risks suffocation or injury. A knotted safety rope hanging near the center of the bin offers little protection against the tremendous pull of unloading grain. Never start machinery before locating children and coworkers.

Disconnect power to the unloading auger before entering the bin. Place an on-off switch at the bin entrance.

Spoiled or wet grain sometimes bridges over the auger and inhibits unloading. Never walk on the bridge because it may collapse and trap you. Break up the bridge with a long pole from the outside.

Wear an effective dust mask when exposed to grain dust. In particular, avoid breathing mold dust from spoiled grain. Handling milo (grain sorghum) is risky, so take precautions against inhaling its dust.

Bunk Planning

Feeding space and bunk planning depend on cattle size and the number that eat at one time. See Table 8-2 and the data summary for space and bunk size requirements.

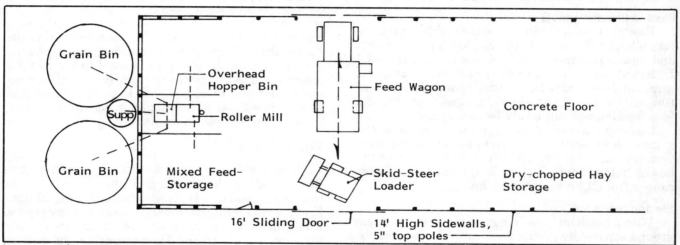

Fig 8-12. Feed handling building.

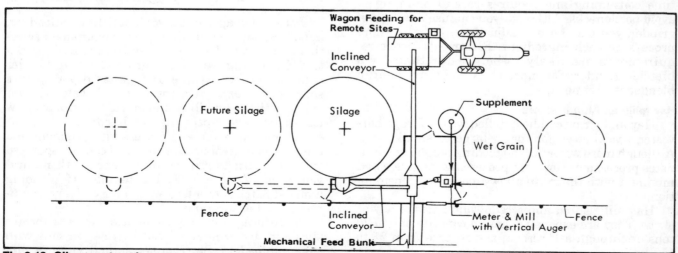

Fig 8-13. Silage wet-grain center, bunk and wagon feeding.
For both fenceline bunks and tightly grouped self feeders or a mechanical conveyor bunk, provide both feed wagon and mechanical conveyor. Meter the ingredients in batches into a stationary mixer over a scale. From the mixer, the feed goes to the feed wagon, self feeders, or a mechanical conveyor bunk. The rest of the system is similar to Fig 8-15.

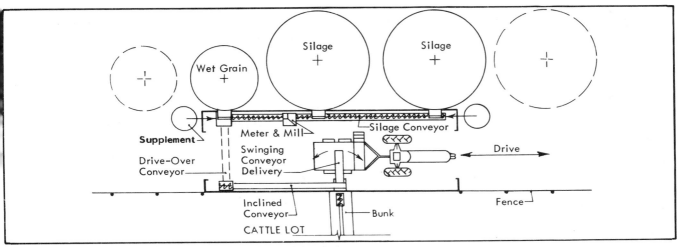

Fig 8-14. Silage grain center, bunk and wagon feeding.
The drive-through building is about 20'x60'.

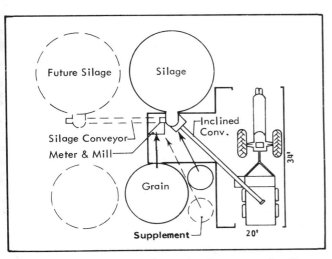

Fig 8-15. Small silage grain center for wagon feeding.
Center building is about 20'x34' plus 12'x16' offset.

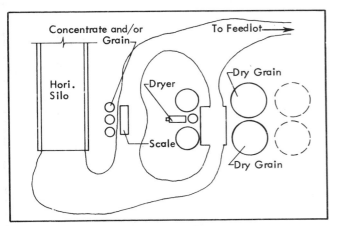

Fig 8-16. Silage dry grain feed layout.
Concentrate bins can be in the dry center. Eliminate the dry grain center if grain is purchased as fed. Wet grain silos can replace the dry bins.

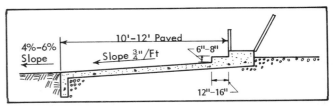

Fig 8-17. Bunk apron.
Use 12' width if the earth below the apron will be muddy or drifted with snow part of the year. A step along the bunk, 6"-8" high by 12"-16" wide, helps reduce cattle manuring in the bunk and reduces damage from scraping equipment.

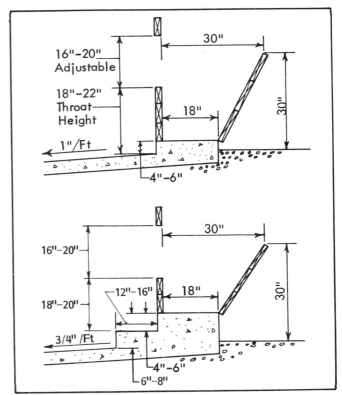

Fig 8-18. Fenceline bunks.

Table 8-2. Bunk design.

Consider drainage, manure buildup, type of feed, and available building materials when selecting a feedbunk. Use a deep bunk for bulky feeds to reduce spillage. Mechanical bunks allow cows to eat from both sides. If a feed conveyor divides the feeding space, make the bunk wider.

Throat height (max.)		**Bunk apron**	
Calves	18″	Slope	¾″-1″/ft
Heifers	20″	Width	10′-12′
Mature cows	24″		
		Neck rails	
Bunk width (max. 60″)		⅜″ cable, 2″ pipe, 2x6 plank	16″-24″ opening
Both sides feeding			
Calves	36″	**Feeding space**	
Heifers	48″-60″	All fed at once	
Mature cows	48″-60″	Calves	18″-22″
One side feeding	18″ bottom width	Heifers	22″-26″
Mechanical feeder	Add 6″-12″ up to max. width	Mature cows	26″-30″
Step along bunk		**Feed always available**	
Height	4″-6″	Hay or silage	4″-6″
Width	12″-16″	Mixed ration	12″-18″

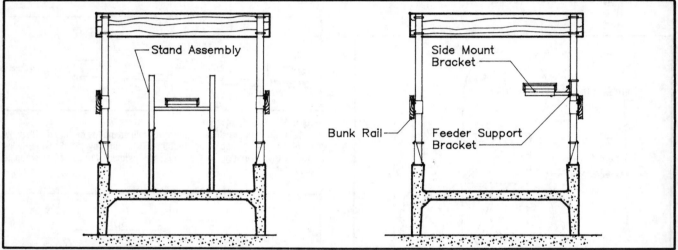

Fig 8-19. Mechanical bunk.

Mechanical bunks are usually wider because cattle eat from both sides. If the feed conveyor divides the eating space, make the bunk wider.

Equipment

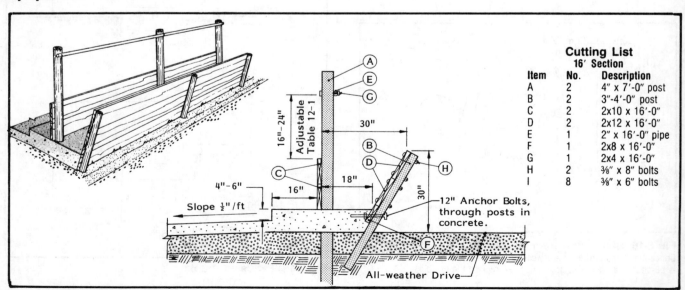

Cutting List
16′ Section

Item	No.	Description
A	2	4″ x 7′-0″ post
B	2	3″-4′-0″ post
C	2	2x10 x 16′-0″
D	2	2x12 x 16′-0″
E	1	2″ x 16′-0″ pipe
F	1	2x8 x 16′-0″
G	1	2x4 x 16′-0″
H	2	⅜″ x 8″ bolts
I	8	⅜″ x 6″ bolts

Fig 8-20. Wooden fenceline bunk.

Posts inside the bunk simplify cleaning along the outside of the bunk and increase bunk capacity.

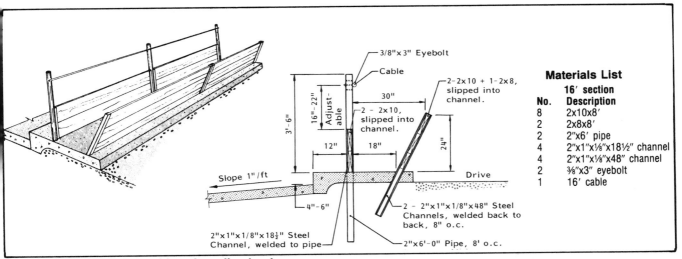

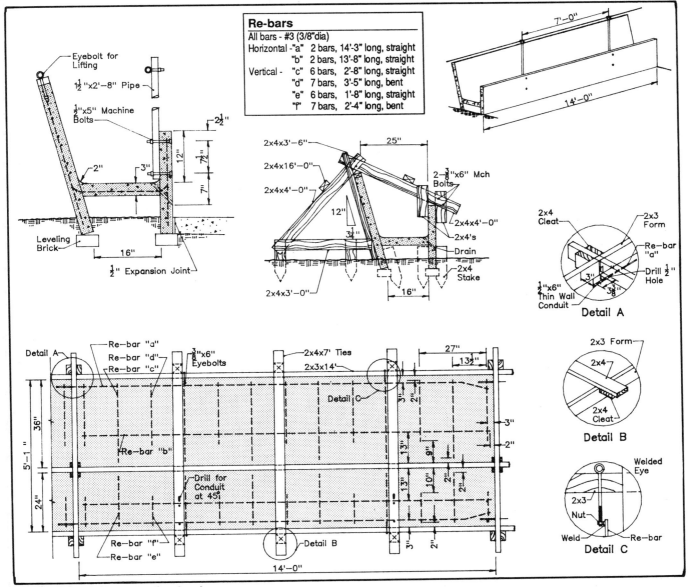

Fig 8-21. Wood and channel iron fenceline bunk.

Materials List
16' section

No.	Description
8	2x10x8'
2	2x8x8'
2	2"x6' pipe
4	2"x1"x1/8"x18½" channel
4	2"x1"x1/8"x48" channel
2	⅜"x3" eyebolt
1	16' cable

Fig 8-22. Concrete fenceline bunk.

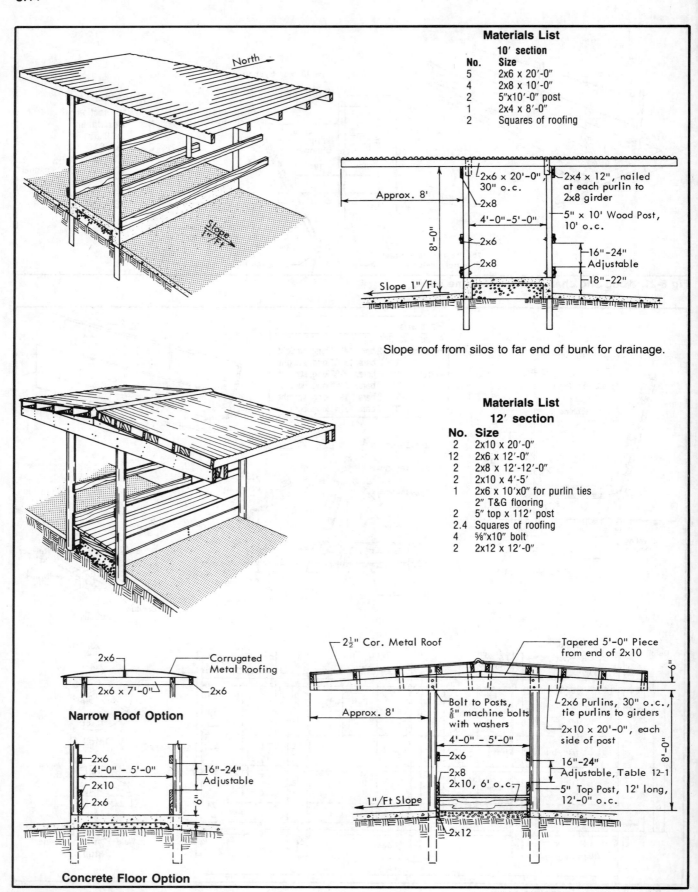

Materials List

10' section

No.	Size
5	2x6 x 20'-0"
4	2x8 x 10'-0"
2	5"x10'-0" post
1	2x4 x 8'-0"
2	Squares of roofing

Approx. 8'

2x6 x 20'-0"
30" o.c.

2x4 x 12", nailed at each purlin to 2x8 girder

2x8

5" x 10' Wood Post, 10' o.c.

4'-0"–5'-0"

8'-0"

2x6

2x8

16"–24"
Adjustable

18"–22"

Slope 1"/Ft

Slope roof from silos to far end of bunk for drainage.

Materials List

12' section

No.	Size
2	2x10 x 20'-0"
12	2x6 x 12'-0"
2	2x8 x 12'-12'-0"
2	2x10 x 4'-5'
1	2x6 x 10'x0" for purlin ties
	2" T&G flooring
2	5" top x 112' post
2.4	Squares of roofing
4	⅝"x10" bolt
2	2x12 x 12'-0"

2½" Cor. Metal Roof

Tapered 5'-0" Piece from end of 2x10

6"

Approx. 8'

Bolt to Posts, ⅝" machine bolts with washers

2x6 Purlins, 30" o.c., tie purlins to girders

4'-0" – 5'-0"

2x10 x 20'-0", each side of post

2x6

2x8

16"–24"
Adjustable, Table 12–1

2x10, 6' o.c.

8'-0"

5" Top Post, 12' long, 12'-0" o.c.

1"/Ft Slope

2x12

Narrow Roof Option

2x6

Corrugated Metal Roofing

2x6 x 7'-0"

2x6

2x6
4'-0" – 5'-0"

2x10

2x6

16"–24"
Adjustable

6"

Concrete Floor Option

Fig 8-23. Covered bunks.
Covered bunks protect feeding equipment and keep moist feed from drying out and dry feed from getting wet. A narrow 6' wide roof provides minimum shade, but a wide roof high enough to clear cleaning equipment can provide adequate summer shade. Orient roofs NNE-SSW to promote thawing and drying around the bunk. Use narrow roof if bunk must be E-W.

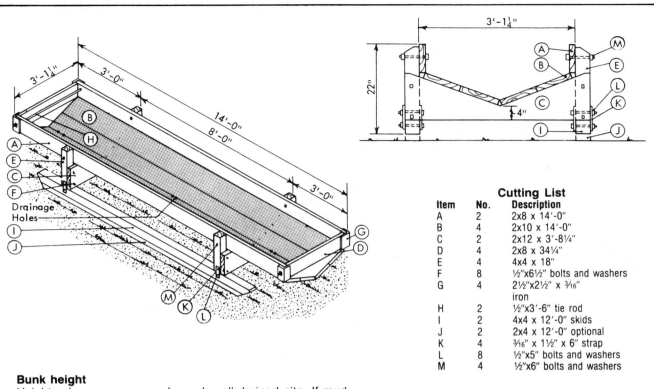

Cutting List

Item	No.	Description
A	2	2x8 x 14'-0"
B	4	2x10 x 14'-0"
C	2	2x12 x 3'-8¼"
D	4	2x8 x 34¼"
E	4	4x4 x 18"
F	8	½"x6½" bolts and washers
G	4	2½"x2½" x ³⁄₁₆" iron
H	2	½"x3'-6" tie rod
I	2	4x4 x 12'-0" skids
J	2	2x4 x 12'-0" optional
K	4	³⁄₁₆" x 1½" x 6" strap
L	8	½"x5" bolts and washers
M	4	½"x6" bolts and washers

Bunk height

Heights shown assume a cleaned, well drained site. If mud, snow, or manure may accumulate, raise bunks to 30" for cows.

Bunk Capacity

Bunks may be up to 60" wide and/or 12" to 14" deep for increased capacity.

Corner Framing

Cutting List

Item	No.	Description
A	2	4x4 x 12'-0"
B	2	2x4 x 12'-0"
C	4	4x4 x 17"
D	2	2x6 x 3'-9½"
E	3	2x12x35½"
F	2	2x12x14'-0"
G	3	2x4x3'-2½"
H	4	2x8x14'-0"
I	1	2x6 x 14'-0"
J	8	½"x6" bolts and washers
K	6	³⁄₈"x4" lag screw
L	4	2½"x2½"x¼" angle iron
M	2	1"x3'-9½" pipe
N	2	½"x3'-6" tie rod
O	2	³⁄₁₆"x1½"x6" strap
P	8	½"x6" bolts and washers

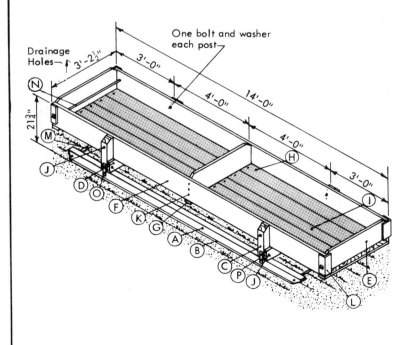

Fig 8-24. Portable lot bunks.
Bunks can be up to 60" wide and/or 12"-14" deep for increased capacity.

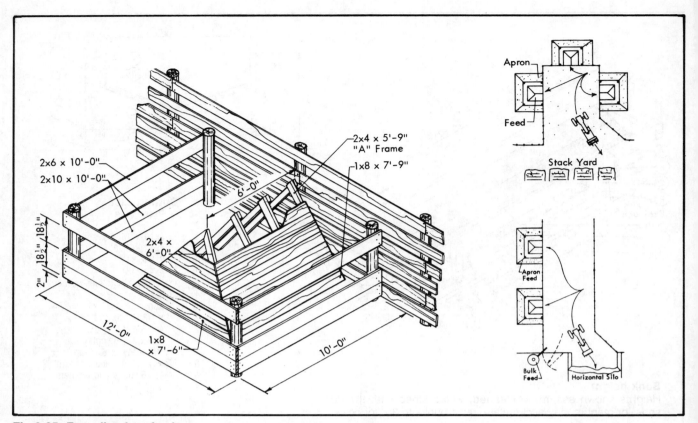

Fig 8-25. Fenceline box feeder.
To make portable, add a plank floor and support skids. More wastage occurs without slanted neckrails.

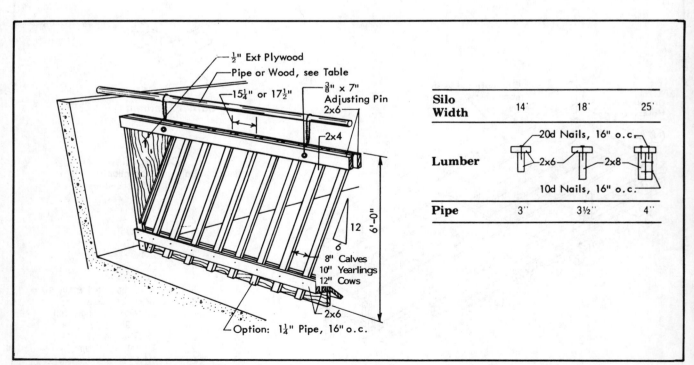

Silo Width	14'	18'	25'
Lumber	2×6	2×8	
Pipe	3"	3½"	4"

Fig 8-26. Horizontal silo feeding fence.

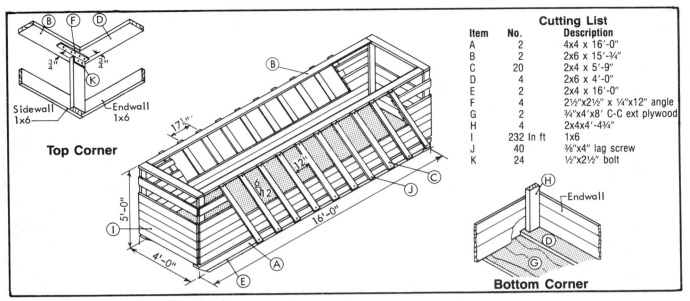

Cutting List

Item	No.	Description
A	2	4x4 x 16'-0"
B	2	2x6 x 15'-¾"
C	20	2x4 x 5'-9"
D	4	2x6 x 4'-0"
E	2	2x4 x 16'-0"
F	4	2½"x2½" x ¼"x12" angle
G	2	¾"x4'x8' C-C ext plywood
H	4	2x4x4'-4¾"
I	232 In ft	1x6
J	40	⅜"x4" lag screw
K	24	½"x2½" bolt

Top Corner

Sidewall 1x6

Endwall 1x6

Bottom Corner

Endwall

Fig 8-27. Wooden hay and silage feeder.

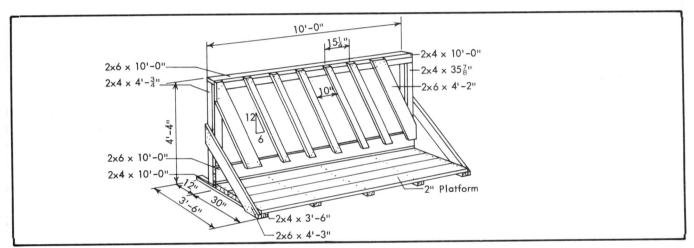

Fig 8-28. Portable feeding panel.

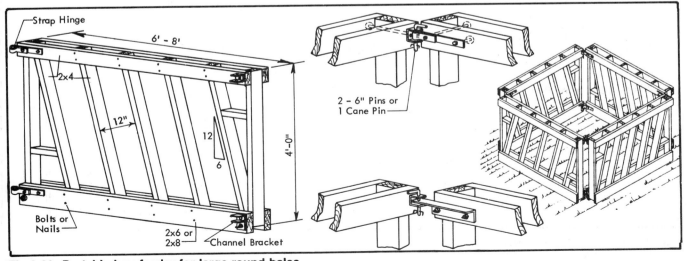

Fig 8-29. Portable box feeder for large round bales.

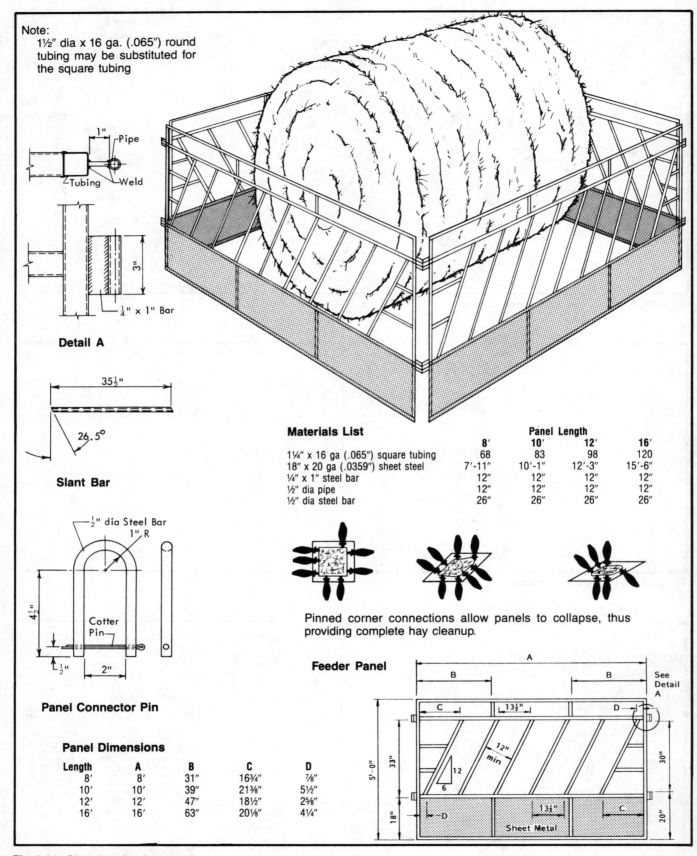

Note:
1½″ dia x 16 ga. (.065″) round tubing may be substituted for the square tubing

Pipe

Weld

Tubing

3″

¼″ x 1″ Bar

Detail A

35½″

26.5°

Slant Bar

½″ dia Steel Bar

1″ R

4½″

Cotter Pin

½″

2″

Panel Connector Pin

Panel Dimensions

Length	A	B	C	D
8′	8′	31″	16¾″	⅞″
10′	10′	39″	21⅜″	5½″
12′	12′	47″	18½″	2⅝″
16′	16′	63″	20⅛″	4¼″

Materials List

	Panel Length			
	8′	10′	12′	16′
1¼″ x 16 ga (.065″) square tubing	68	83	98	120
18″ x 20 ga (.0359″) sheet steel	7′-11″	10′-1″	12′-3″	15′-6″
¼″ x 1″ steel bar	12″	12″	12″	12″
½″ dia pipe	12″	12″	12″	12″
½″ dia steel bar	26″	26″	26″	26″

Pinned corner connections allow panels to collapse, thus providing complete hay cleanup.

Feeder Panel

A

B

B

See Detail A

C

13½″

D

5′-0″

33″

12″ min

12

6

30″

18″

D

13½″

C

Sheet Metal

20″

Fig 8-30. Slant bar feeder panels.
For pole barns or to feed stacks on pasture.

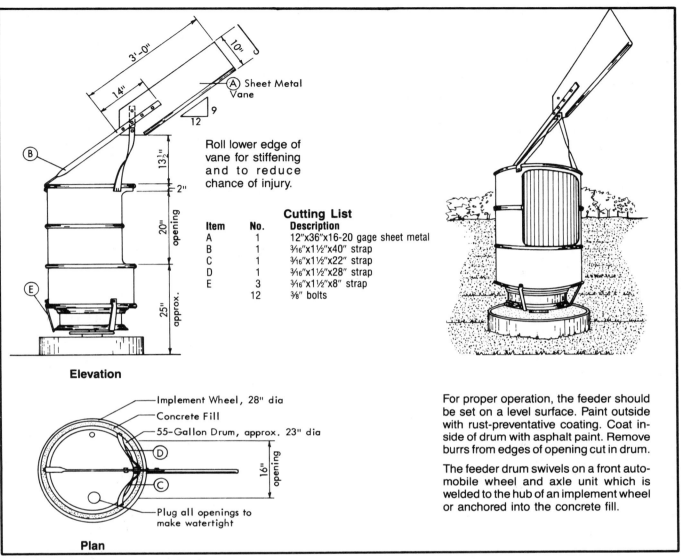

Cutting List

Item	No.	Description
A	1	12"x36"x16-20 gage sheet metal
B	1	3/16"x1½"x40" strap
C	1	3/16"x1½"x22" strap
D	1	3/16"x1½"x28" strap
E	3	3/16"x1½"x8" strap
	12	3/8" bolts

Roll lower edge of vane for stiffening and to reduce chance of injury.

Elevation

- Implement Wheel, 28" dia
- Concrete Fill
- 55-Gallon Drum, approx. 23" dia
- Plug all openings to make watertight

Plan

For proper operation, the feeder should be set on a level surface. Paint outside with rust-preventative coating. Coat inside of drum with asphalt paint. Remove burrs from edges of opening cut in drum.

The feeder drum swivels on a front automobile wheel and axle unit which is welded to the hub of an implement wheel or anchored into the concrete fill.

Fig 8-31. Weathervane mineral feeder.

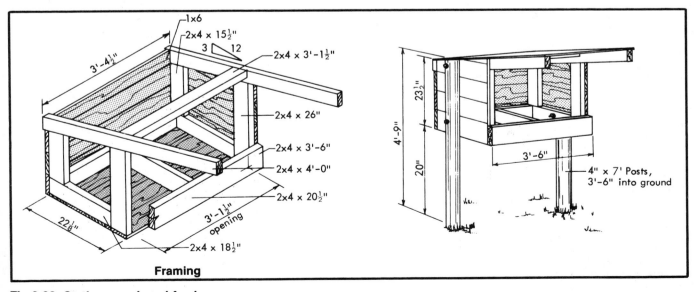

Framing

Fig 8-32. Stationary mineral feeder.

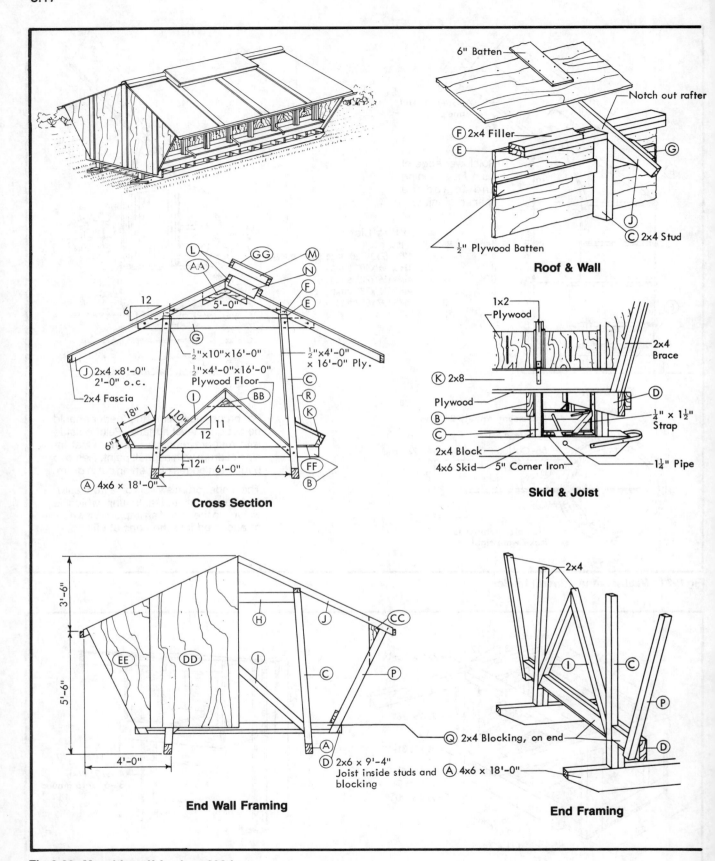

Roof & Wall

6" Batten

Notch out rafter

(F) 2x4 Filler

(E)

(G)

(J)

(C) 2x4 Stud

½" Plywood Batten

Cross Section

(L) (AA) (GG) (M) (N) (F) (E)

12
6

5'-0"

(G)
½"x10"x16'-0"

½"x4'-0"
x 16'-0" Ply.

(J) 2x4 x8'-0"
2'-0" o.c.

½"x4'-0"x16'-0"
Plywood Floor

(C)

2x4 Fascia

18"

10"

(I) (BB) (R) (K)

11
12

6"

12"

6'-0"

(FF)

(A) 4x6 x 18'-0"

(B)

Skid & Joist

1x2
Plywood

2x4
Brace

(K) 2x8

(D)

Plywood

(B)
(C)

2x4 Block

4x6 Skid

5" Corner Iron

¼" x 1½"
Strap

1¼" Pipe

End Wall Framing

3'-6"

5'-6"

(H) (J) (CC)

(EE) (DD) (I)

(C) (P)

(A)

4'-0"

(D) 2x6 x 9'-4"
Joist inside studs and
blocking

End Framing

2x4

(I) (C)

(P)

(D)

(Q) 2x4 Blocking, on end

(A) 4x6 x 18'-0"

Fig 8-33. Movable self feeder—300 bu.
Capacity for about 100 yearlings.

CUTTING LIST

Item	No.	Description
A	2	4x6 x 18'-0'' skid
B	7	2x6 x 8'-6'', 2'-0'' o.c.
C	18	2x4 x 7'-0'' x 2'-0'' o.c.
D	2	2x6 x 9'-4'' end joist
E	2	2x4 x 15'-11''
F	4	2x4 x 22''
	12	2x4 x 22½''
G	7	2x6 x 8'-0''
H	2	2x6 x 5'-0''
I	18	2x4 x 4'-4'', 2'-0'' o.c.
J	18	2x4 x 8'-0'', 2'-0'' o.c. rafter
K	2	2x8 x 15'-11''
L	2	2x4 x 7'-8½''
	2	2x4 x 7'-9''
M	2	2x4 x 24''
N	2	2x6 x 17½''
P	4	2x4 x 5'-3''
Q	2	2x4 x 8'-0'' cut for blocking
R	6	¼''x1''x28'' strap, 4'-0'' o.c.
	6	2x2 x 16⅜''
S	32	1x2 x 24''
	8	Sheets plywood; see cutting diagram
	21	Sheets ½'' x 4' x 8' exterior plywood batten, door, roofing, wall, floor

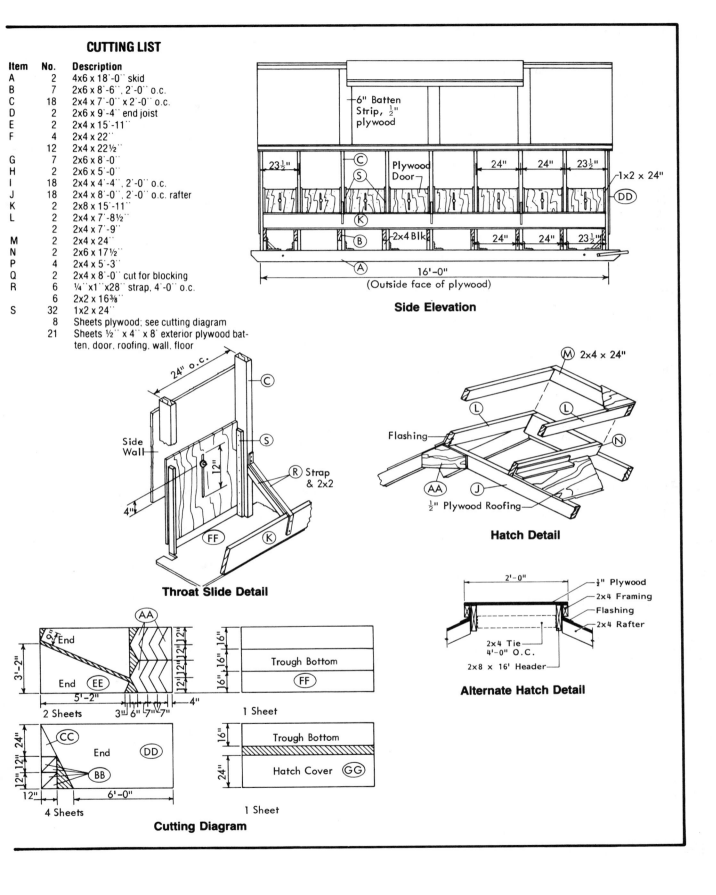

Side Elevation

Throat Slide Detail

Hatch Detail

Alternate Hatch Detail

Cutting Diagram

Silo Capacities and Feed Data

Table 8-3. Horizontal silo capacity, wet tons.
65% moisture; 40 lb/ft^3 or 50 ft^3 = 1 ton; 1.25 ft^3/bu. Silo assumed level full. Capacities rounded to nearest 5 tons. To calculate capacity of other silo sizes: (silage depth, ft × silo width, ft × silo length, ft) ÷ 50.

Depth ft	Silo floor width, ft								
	20	30	40	50	60	70	80	90	100
	- - - - - - - - - - wet tons/10' length - - - - - - - - -								
10	40	60	80	100	120	140	160	180	200
12	50	70	95	120	145	170	190	215	240
14	55	85	110	140	170	195	225	250	280
16	65	95	130	160	190	225	255	290	320
18	70	110	145	180	215	250	290	325	360
20	80	120	160	200	240	280	320	360	400

Table 8-4. Horizontal silo capacity, dry matter.
Silo assumed level full. Capacities rounded to nearest 5 tons.

Depth ft	Silo floor width, ft								
	20	30	40	50	60	70	80	90	100
	- - - - - - - - - tons dry matter/10' length - - - - - - - -								
10	15	20	30	35	40	50	55	65	70
12	15	25	35	40	50	60	65	75	85
14	20	30	40	50	60	70	80	90	100
16	20	35	45	55	65	80	90	100	110
18	25	40	50	65	75	90	100	115	125
20	30	40	55	70	85	100	110	125	140

Table 8-5. Maximum exposed surface of horizontal silos.
Table values based on removing a 4″ slice/day to reduce spoilage.

Feeding rate lb/animal	Surface area ft²/animal
20	2
30	3
40	4
50	5
60	6
70	7
80	8

Table 8-10. Silage removal.
Amount of silage removal to reduce spoilage.

Silage type	Silage removed, in/day Weather	
	Cold	Hot
Whole corn	2	4-6
Alfalfa-brome	2	3-4
Chopped ear corn	2	2
Cracked shell corn	4	4

Table 8-11. Dry matter factor (DMF).
To convert from wet tons to tons of dry matter, divide by the DMF. To convert from tons of dry matter to wet tons, multiply by the DMF. DMF = 100 ÷ (100 − % moisture).

% moisture	DMF
30	1.43
40	1.67
50	2.00
55	2.22
60	2.50
65	2.86
70	3.33
75	4.00
80	5.00

Table 8-12. Grain moisture factor (GMF).
To convert wet tons to tons at 15.5% moisture, divide by GMF. To convert tons at 15.5% moisture to wet tons, multiply by GMF. GMF = 84.5 ÷ (100 − % moisture).

% moisture	GMF
18	1.03
20	1.06
22	1.08
24	1.11
26	1.14
28	1.17
30	1.21
32	1.24
34	1.28
36	1.32
38	1.36
40	1.41
45	1.54
50	1.69

Table 8-6. Tower silo capacity, wet tons.

Silage at 65% moisture, wet basis. Capacities allow 1' of unused depth for settling in silos up to 30' high and 1' more for each 10' beyond 30' height. Capacities rounded to nearest 5 tons. Adapted from *1983 ASAE Yearbook*.

Silo height ft	Silo diameter, ft								
	14	16	18	20	22	24	26	28	30
	- - - - - - - - - - - - - wet tons - - - - - - - - - - - - -								
20	45	60	75	95	115	135	160	185	210
24	60	75	95	125	150	175	205	235	275
28	75	100	125	150	185	215	255	295	340
32	90	115	150	185	225	265	310	365	415
36	105	135	175	215	265	310	370	430	490
40	125	165	205	255	305	365	430	495	570
44	145	185	235	290	350	420	490	570	655
48	160	210	265	330	400	475	560	645	745
52	185	235	300	370	450	530	625	725	830
56	205	265	335	410	500	590	695	805	925
60	225	290	370	455	550	650	780	885	1020
64			405	500	600	715	850	970	1120
68			445	545	650	780	925	1060	1215
72						840	1000	1145	1310
76						890	1075	1220	1400
80						955	1120	1300	1485
84							1215	1385	1575
88							1285	1475	1660
92							1375	1560	1745
96							1445	1640	1830
100							1515	1730	1930

Table 8-7. Tower silo capacity, tons dry matter.

Capacities allow 1' of unused depth for settling in silos up to 30' high and 1' more for each 10' beyond 30' height. To determine silo capacity in "wet tons," multiply silo capacity value, dry matter, by the DMF value from Table 8-11. Capacities rounded to nearest 5 tons. Adapted from *1983 ASAE Yearbook*.

Silo height ft	Silo diameter, ft								
	14	16	18	20	22	24	26	28	30
	- - - - - - - - - - - tons of dry matter - - - - - - - - - - -								
20	15	20	25	35	40	45	55	65	75
24	20	25	35	45	50	60	70	85	95
28	25	35	45	55	65	75	90	105	120
32	30	40	50	65	80	95	110	125	145
36	35	50	60	75	90	110	130	150	170
40	45	55	70	90	105	125	150	175	200
44	50	65	80	100	125	145	170	200	230
48	55	75	95	115	140	165	195	225	260
52	65	85	105	130	155	185	220	255	290
56	70	95	115	145	175	205	245	280	325
60	80	100	130	160	190	230	275	310	355
64			140	175	210	250	300	340	390
68			155	190	230	270	325	370	425
72						295	350	400	460
76						315	375	425	490
80						335	400	455	520
84							425	485	550
88							450	515	580
92							480	545	610
96							505	574	640
100							530	605	675

Table 8-8. Tower silo capacity for ground ear and whole shelled corn.

For 30% moisture shelled corn and 27% kernel moisture (32% blend moisture) ear corn and 60 lb/bu (1.25 ft³/bu) density. Capacities allow for 1' of unused depth and are rounded to the nearest 5 tons and 10 bu. Capacities increase approximately 5% with moisture contents between 25%-35%. Capacities also vary with fineness of grind and filling procedure. DM = dry matter. Adapted from *Penn State University Special Circular No. 143*.

Silo height ft	Unit of measure	Silo diameter, ft							
		10	12	14	16	17	18	20	22
20	Wet tons	35	50	70	95				
	Tons DM	25	35	50	65				
	Bushels	1190	1720	2340	3060				
24	Wet tons	45	65	85	110	125			
	Tons DM	30	45	60	75	85			
	Bushels	1450	2080	2830	3700	4180			
28	Wet tons	50	75	100	130	150	165	205	
	Tons DM	35	50	70	90	100	115	140	
	Bushels	1700	2440	3330	4340	4900	5500	6790	
32	Wet tons	60	85	115	150	170	190	235	290
	Tons DM	40	55	80	100	115	130	160	195
	Bushels	1950	2800	3820	4990	5630	6310	7790	9430
36	Wet tons		95	130	170	190	215	265	325
	Tons DM		65	90	115	130	145	180	220
	Bushels		3170	4310	5630	6360	7130	8800	10640
40	Wet tons		105	145	190	210	240	295	365
	Tons DM		70	100	130	145	160	200	250
	Bushels		3530	4800	6270	7080	7940	9800	11860
44	Wet tons			160	205	235	265	325	400
	Tons DM			110	140	160	180	220	270
	Bushels			5300	6920	7810	8750	10810	13080
48	Wet tons			175	225	255	285	355	435
	Tons DM			120	155	175	195	240	295
	Bushels			5790	7560	8530	9570	11810	14290
52	Wet tons				245	280	310	385	475
	Tons DM				165	190	210	260	325
	Bushels				8200	9260	10380	12820	15510
56	Wet tons				265	300	335	415	510
	Tons DM				180	205	230	285	345
	Bushels				8850	9990	11200	13820	16730
60	Wet tons				285	325	365	450	545
	Tons DM				195	220	245	305	370
	Bushels				9490	10710	12010	14830	17940

Table 8-9. Tower silo capacity for ground shelled corn.

For 30% moisture grain and 68.0 lb/bu (1.25 ft³/bu) density. Capacities allow for 1' of unused depth and are rounded to the nearest 5 tons and 10 bu. Capacity increases 5% with moisture contents between 25%-35%. Capacity also varies with fineness of grind and filling procedures. DM = dry matter. Adapted from *Penn State University Special Circular No. 143.*

| Silo height ft | Unit of measure | Silo diameter, ft | | | | | | | |
		10	12	14	16	17	18	20	22
20	Wet tons	40	60	80	105				
	Tons DM	30	40	55	75				
	Bushels	1190	1720	2340	3060				
24	Wet tons	50	70	100	130	145			
	Tons DM	35	50	70	90	100			
	Bushels	1450	2080	2830	3700	4180			
28	Wet tons	60	85	115	150	170	190	235	
	Tons DM	40	60	80	105	120	130	165	
	Bushels	1700	2440	3330	4340	4900	5500	6790	
32	Wet tons	65	95	130	170	195	215	265	
	Tons DM	45	65	90	120	135	150	185	
	Bushels	1950	2800	3820	4990	5630	6310	7790	9430
36	Wet tons		110	150	195	220	245	300	375
	Tons DM		75	105	135	150	170	210	260
	Bushels		3170	4310	5630	6360	7130	8800	10640
40	Wet tons		120	165	215	240	270	335	415
	Tons DM		85	115	150	170	190	260	320
	Bushels		3530	4800	6270	7080	7940	9800	11860
44	Wet tons			180	235	265	300	370	455
	Tons DM			125	165	185	210	260	320
	Bushels			5300	6920	7810	8750	10810	13080
48	Wet tons			200	260	290	325	405	495
	Tons DM			140	180	205	230	280	345
	Bushels			5790	7560	8530	9570	11810	14290
52	Wet tons				280	315	355	440	535
	Tons DM				195	220	250	305	375
	Bushels				8200	9260	10380	12820	15510
56	Wet tons				305	340	385	475	575
	Tons DM				210	240	270	330	400
	Bushels				8850	9990	11200	13820	16730
60	Wet tons				325	370	415	510	620
	Tons DM				230	260	290	355	435
	Bushels				9490	10710	12010	14830	17940

Table 8-13. Grain densities.

For grain moisture content cited.

| | Approximate weight | |
	lb/bu	lb/ft³
Barley, 15%	48	38.4
Corn, 15½%		
Ear, husked	70	28
Shelled	56	44.8
Flaxseed, 11%	56	44.8
Grain sorghum, 15%	56 & 50	44.8 & 40
Oat, 16%	32	25.6
Rye, 16%	56	44.8
Soybeans, 14%	60	48
Wheat, 14%	60	48

Table 8-15. Storage capacity for round grain bins.

Capacity does not include space above eave line. Based on 1.25 ft³/bushel.

| Dia. ft | Depth of grain, ft | | | | |
	1	11	13	16	19
			bushels		
14	125	1,375	1,625	2,000	2,375
18	203	2,200	2,635	3,250	3,850
21	277	3,050	3,600	4,400	5,300
24	362	4,000	4,700	5,800	6,900
27	458	5,050	5,950	7,300	8,700
30	565	6,215	7,345	9,040	10,735
36	814	8,950	10,600	13,000	15,450
40	1,005	11,050	13,050	16,100	19,100

Table 8-14. Approximate capacity of round, hopper-bottom bins.
60° hopper; 24″ slide valve clearance.

Description	Overall height ft	Capacity in tons lb/ft³ material			Total capacity	
		30	40	50	ft³	bu
6′ diameter center draw-off	10-10½	2.2	2.7	3.4	135	108
	10½-13	3.1	4.2	5.3	210	166
	15½-16	4.2	5.7	7.1	285	228
	18-18½	5.4	7.2	9.0	360	288
	20-20½	6.2	8.6	10.4	415	322
6′ diameter side draw-off	14½-15	2.8	3.7	4.7	187	150
	17-17½	3.9	5.2	6.6	263	210
	19½-20	5.0	6.7	8.4	338	270
	22½-23	6.1	8.2	10.3	413	330
	25-25½	7.3	9.7	12.1	487	390
9′ diameter center draw-off	16½-17	8.4	11.2	14.0	561	413
	19½-20	11.0	14.6	18.3	730	583
	22-22½	13.5	18.0	22.5	900	720
	24½-25	16.0	21.3	26.7	1,067	853
	27½-28	18.6	24.7	30.9	1,236	990
12′ diameter center draw-off	20-20½	16.3	21.7	27.1	1,085	870
	22½-23	20.7	27.7	34.6	1,383	1,110
	25½-26	25.2	33.6	42.0	1,681	1,345
	28-28½	29.7	39.6	49.5	1,980	1,585
	30-30½	34.2	45.6	57.0	2,278	1,820
	33-33½	38.7	51.5	64.4	2,577	2,060
	36-36½	43.1	57.5	71.9	2,875	2,300
	38½-39	47.6	63.5	79.3	3,174	2,540
	41½-42	52.0	69.4	86.8	2,472	2,780

Table 8-16. Approximate capacity of ear corn cribs.
Based on 2½ ft³/bushel. Round cribs include ½ cone space under 1:1 roof slope but no deduction for center tunnel.

Rectangular			Round		
Width ft	Height ft	Bu per 10 ft length	Dia. ft	Height ft	Bu
4	12	188	12	12	540
	16	256		16	720
	20	320		20	900
6	12	288	14	12	740
	16	384		16	980
	20	480		20	1,230
8	12	384	16	12	960
	16	512		16	1,280
	20	638		20	1,610
10	12	480	18	12	1,220
	16	640		16	1,620
	20	800		20	2,030

Table 8-17. Hay and straw densities.

Material	ft³/ton	lb/ft³
Loose		
Alfalfa	450-500	4-4.4
Non legume	450-600	3.3-4.4
Straw	670-1,000	2-3
Baled		
Alfalfa	200-330	6-10
Non legume	250-330	6-8
Straw	400-500	4-5
Chopped		
Alfalfa, 1½″ cut	285-360	5.5-7
Non legume, 3″ cut	300-400	5-6.7
Straw	250-350	5.7-8

Table 8-18. Hay shed capacities.
Shed has 20′ high sidewalls.

Shed width, ft	Shed capacity		
	Baled	Chopped	Loose
		tons/ft of length	
24	2.0	1.9	0.8
30	2.6	2.3	1.0
36	3.1	2.8	1.2
40	3.4	3.1	1.4
48	4.0	3.7	1.7

9. WATER AND WATERERS

Water Supply

Water is needed for drinking, sanitation, cooling sprays, waterer spillage, and fire protection. Total water needs can be high. Water consumption depends on animal size, activity, diet, and season of the year. Needs range from 9 gal/1,000 lb per day during winter to 18 gal/1,000 lb per day during hot weather. Fire protection requires a lot of water quickly. Stockpile a 2 hr supply of at least 10 gallons per minute (gpm)—preferably 50 gpm—for fire protection. Size the water supply system to meet maximum needs.

For pasture or range systems, provide water tanks with at least a 1-day supply. Because range cattle usually drink all at once, 1 or 2 times/day, size the pump and pipelines to provide a day's water in 4 hr. Feedlot watering systems require tanks with at least 50% of a 1-day supply available. Size the pump and pipelines to provide a day's water in 8 hr. Protect large operations against pump failure with a gravity flow storage system with at least a 1-day supply.

Consider friction losses in water lines, operating pressure, pressure head due to well depth, and elevation differences when sizing pumps. Water use is intermittent and fluctuates, so install a pressure tank or storage reservoir. A pressure tank with 10 to 12 min pumping capacity or a reservoir with at least one day's water supply is recommended. Multiple pressure tanks are preferred over one large tank in case of a malfunction. Connect pressure tanks in parallel to reduce backpressure on the pump during peak flow periods, Fig 9-1.

Consider intermediate water storage, especially if standby power is not available. With low yield wells, a small pump can stockpile water during nonpeak periods.

ated waterers for a fresh supply of water and continuous overflow waterers for pastures.

Plumbing in livestock facilities encounters special corrosion problems. PVC, CPVC, or polyethylene pipe with nylon fittings and stainless steel fasteners improve corrosion resistance. Check with the manufacturer for the proper temperature and pressure rating of each type of pipe before installing.

Overhead pipes in naturally ventilated buildings require heat tape and insulation. Insulation alone usually will not prevent freezing in open-front buildings. Drain water systems in unused buildings. Use antifreeze to protect traps that cannot be drained.

Buried lines reduce freezing problems but are difficult to repair. Frost penetration in the Midwest varies from 3' to over 6'. In compact soil, such as under driveways and animal traffic areas, and in areas where snow cover is blown away or removed, the frost depth can increase by 2'. Bring underground pipes to the surface through a 6"-12" diameter rigid plastic pipe to protect them from damage. Extend the pipe below the frost level to help prevent freezing. Frost free waterers are commercially available.

Heavily insulated nonheated stock waterers are also available. They generally remain ice free if water entering them is above 40 F and at least one tank full of water is consumed every 4 to 6 hr. Their success depends on the design and weather conditions.

Prevent all waterers from back siphoning. Use either an antiback siphoning device in the water line or an air gap between the water inlet and the maximum water level. Include antiback siphoning devices on all hose bibs. Waterers must be easy to drain, clean, and maintain. See MWPS-14, *Private Water Systems Handbook,* for information on design of pipes, pumps, and water supplies.

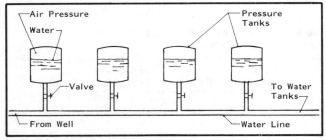

Fig 9-1. Pressure tank layout.

Waterers

Provide one watering space for each 25 head (1 cup or 2 linear ft of tank perimeter). A minimum of two waterer locations per group of animals is recommended. Waterers located in a fenceline can serve two lots and help new animals find water as they walk the fence, but limits accessibility. Waterers in the lot are accessible by more cattle and reduce the chance of boss animals keeping timid animals away from water. Water depths of 6"-8" are preferred. Use deeper tanks where supply capacity is limited. Consider float-oper-

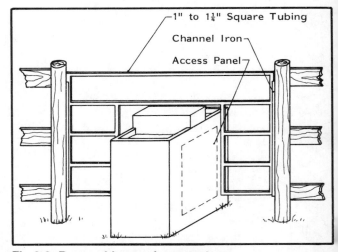

Fig 9-2. Removable panel over waterer.

Installation

Install waterers on concrete slabs. Provide a rough surface to minimize slipping—a self draining

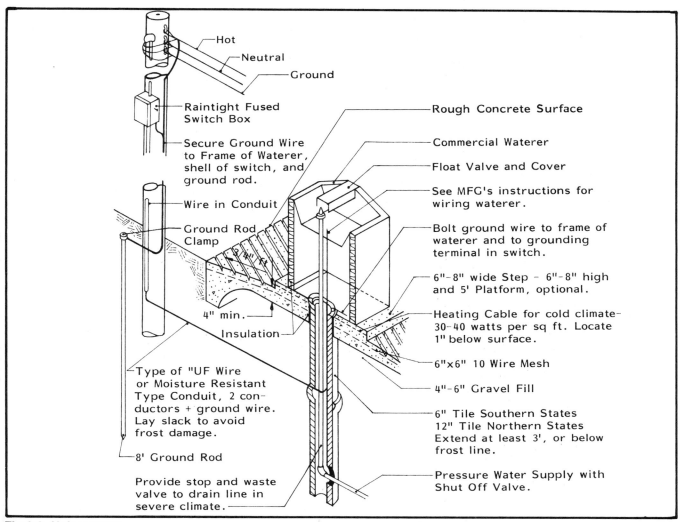

Fig 9-3. Unit waterer.

Labels in figure:
- Hot
- Neutral
- Ground
- Raintight Fused Switch Box
- Secure Ground Wire to Frame of Waterer, shell of switch, and ground rod.
- Wire in Conduit
- Ground Rod Clamp
- 4" min.
- Insulation
- Type of "UF Wire or Moisture Resistant Type Conduit, 2 conductors + ground wire. Lay slack to avoid frost damage.
- 8' Ground Rod
- Provide stop and waste valve to drain line in severe climate.
- Rough Concrete Surface
- Commercial Waterer
- Float Valve and Cover
- See MFG's instructions for wiring waterer.
- Bolt ground wire to frame of waterer and to grounding terminal in switch.
- 6"-8" wide Step – 6"-8" high and 5' Platform, optional.
- Heating Cable for cold climate– 30-40 watts per sq ft. Locate 1" below surface.
- 6"x6" 10 Wire Mesh
- 4"-6" Gravel Fill
- 6" Tile Southern States 12" Tile Northern States Extend at least 3', or below frost line.
- Pressure Water Supply with Shut Off Valve.

diamond pattern is preferred. Extend the slab away from the waterer at least 10' in each direction. Slope concrete ¾"/ft away from the waterer for good drainage. Follow the general guidelines in Fig 9-3 and the manufacturer's recommendations for installing automatic waterers.

Heater mats can be effective for melting snow and ice on concrete slabs, Fig 9-4. Provide a watt density of 10 to 20 watts/ft². Install the mat 1" below the concrete surface. Control the mat manually with a switch and pilot light indicator. When ice begins to accumulate, turn on the mat for 1 to 2 hr and remove the loosened ice buildup.

Install a fused disconnect switch for each electrically heated waterer. Use a waterproof switch and corrosion resistant box with appropriate fittings. Fuse only the "hot" wire with a fuse 25% larger than the total amperage rating of the waterer.

Enclose all wires mounted on a service pole in nonmetallic conduit. Seal the top of the conduit or extend it into a weatherhead to keep water out. Extend the conduit at least 24" into the ground. Install all switches and other electrical equipment where

they cannot be damaged. Careful sizing of wiring for waterers is important for reliable and safe operations. All electrically heated waterers should have a grounding conductor as part of the electrical cable. A ground rod alone does not provide adequate protection.

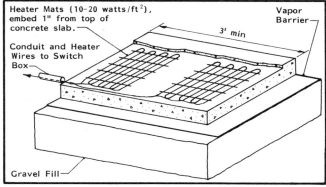

Labels in figure:
- Heater Mats (10-20 watts/ft²), embed 1" from top of concrete slab.
- Conduit and Heater Wires to Switch Box
- 3' min
- Vapor Barrier
- Gravel Fill

Fig 9-4. Heater mat installation.

Preventing Freezing

Water supplies for cattle are often at remote sites where accessibility during freezing weather is difficult. The greatest heat loss from water tanks is evaporation, which is enhanced by wind. Possible heat sources for preventing water freezing are:

• Heat from the ground.
• Heat from incoming water.
• Supplemental heat sources.

Windbreak fences or shelterbelts are effective in reducing wind velocities and heat loss. Provide cattle access on the south side. See the farmstead planning chapter for windbreak details.

Circulating standing water can be used to reduce freezing. Circulating devices bring the warmer water up to the surface to thaw a portion of the ice layer. Propane gas bubblers and submerged propellers are devices to circulate water.

A propane gas bubbler slowly releases pressurized gas near the bottom to circulate the lower warmer water to the surface, Fig 9-5. This maintains an ice free area of about 1½'-3' in diameter. A submerged propeller circulates warmer water to the surface to maintain an ice free area, Fig 9-6. These propellers work well in open ponds but are not recommended for stock tanks. They have been used in stock tanks that are insulated and partially covered. This system depends on wind to be effective. With no wind, the tank freezes.

Another way to reduce freezing is to insulate the waterer and provide only a small access area for livestock. This exposes only a small portion of the water surface to the freezing temperatures and uses heat from incoming water and the ground to keep water from freezing. Mass insulation is one of the simplest and most direct approaches to preventing waterer freezing.

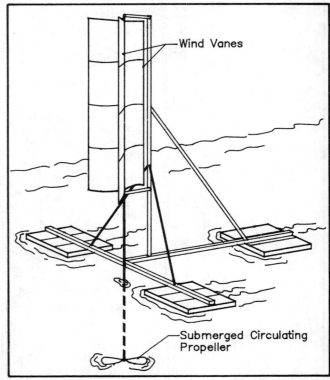

Fig 9-6. **Floating, wind driven, vertical axis driveshaft unit.**

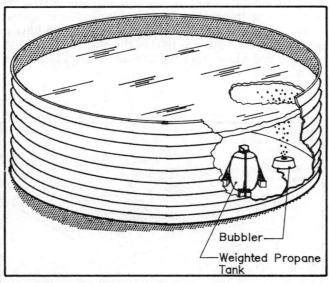

Fig 9-5. **Propane gas bubbler.**

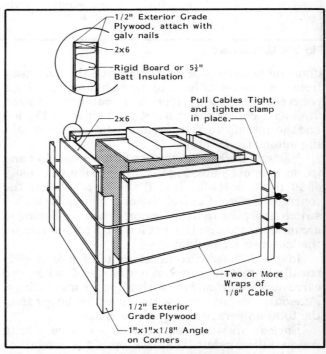

Fig 9-7. **Super insulated automatic waterer.**

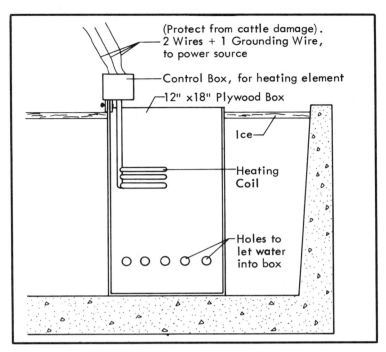

Fig 9-8. Box for stock tank heater.
Install box around inlet and heater to reduce volume of water to be heated.

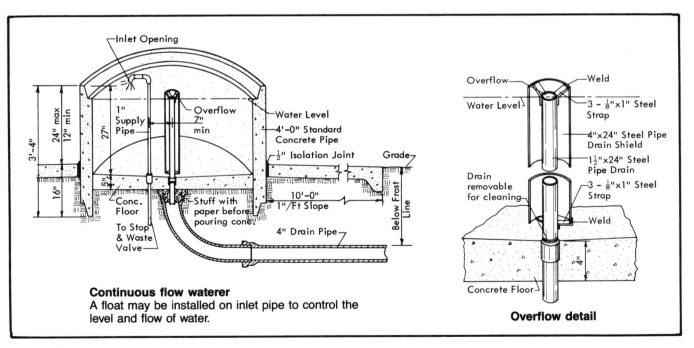

Continuous flow waterer
A float may be installed on inlet pipe to control the
level and flow of water.

Overflow detail

Fig 9-9. Continuous flow waterer.

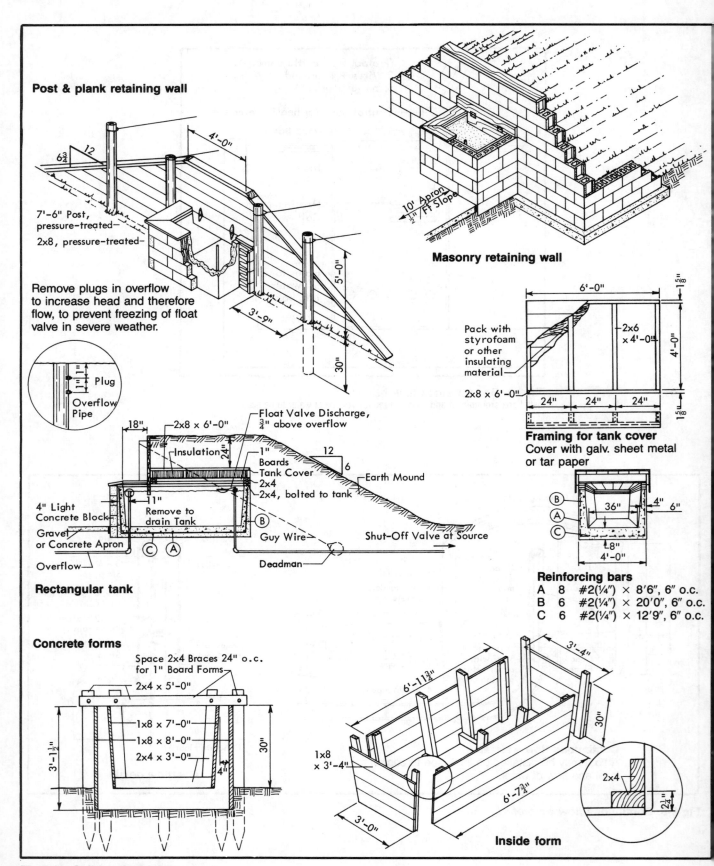

Post & plank retaining wall

6¾ · 12

4'-0"

7'-6" Post, pressure-treated

2x8, pressure-treated

5'-0"

3'-9"

30"

Remove plugs in overflow to increase head and therefore flow, to prevent freezing of float valve in severe weather.

Plug
Overflow Pipe

Masonry retaining wall

10' Apron ½/Ft Slope

6'-0"

2x6 × 4'-0"

Pack with styrofoam or other insulating material

2x8 × 6'-0"

24" 24" 24"

4'-0"

1⅝"

1⅝"

Framing for tank cover
Cover with galv. sheet metal or tar paper

18" 2x8 × 6'-0" Float Valve Discharge, ¾" above overflow

Insulation

24"

1" Boards
Tank Cover
2x4
2x4, bolted to tank

12 · 6

Earth Mound

11"

4" Light Concrete Block

Remove to drain Tank

B

Gravel or Concrete Apron

C A

Guy Wire

Shut-Off Valve at Source

Overflow

Deadman

Rectangular tank

B
A
C

36"

4" 6"

8"
4'-0"

Reinforcing bars

A 8 #2(¼") × 8'6", 6" o.c.
B 6 #2(¼") × 20'0", 6" o.c.
C 6 #2(¼") × 12'9", 6" o.c.

Concrete forms

Space 2x4 Braces 24" o.c. for 1" Board Forms
2x4 × 5'-0"
1x8 × 7'-0"
1x8 × 8'-0"
2x4 × 3'-0"

3'-1½"

30"

4"

3'-4"

6'-11¾"

30"

1x8 × 3'-4"

6'-7¾"

3'-0"

2x4

2¼"

Inside form

Fig 9-10. Soil insulated waterer.

10. UTILITIES

Utilities for beef facilities include electricity for feed handling, lighting, heating, and ventilation; gas for heating; and water. Check with your insurance company and local building official, where applicable, for approved practices before making major changes or installing new utilities.

Electrical

This section outlines general materials and methods for electrical equipment and wiring in beef facilities. It does **not** cover wiring from the power supplier to the buildings or sizing of building distribution panels and circuits. Electrical power requirements vary greatly depending on the size of the herd and electrical equipment used, such as silo unloaders and feed mixing/milling equipment. For more information about electrical wiring for farm buildings, get MWPS-28, *Farm Buildings Wiring Handbook,* from one of the addresses listed inside the front cover of this book.

To ensure proper wiring, consult:
• Your local power supplier to help plan and install the distribution system.
• A licensed electrician to help plan and install distribution panels and motor circuits, select conductors and fixtures, and verify compliance with state and local codes.
• Your electrical equipment supplier for dust- and moisture-tight fixtures and wiring required for damp buildings. Plan ahead because this equipment is only available through electrical wholesale supply stores. You may need to order the equipment.
• Your insurance company. If the electrical system does not meet their standards, they may increase rates or refuse to insure the building.
• Local building officials where applicable.

Materials

Use electrical materials that have been safety tested by an approved agency such as the Underwriters' Laboratory (UL). This includes switches, plugs, connectors, and receptacles. The Underwriters' Laboratories (UL) is a non-profit organization that tests and establishes standards for various products. The UL approves items as having met minimum safety standards and lists them subject to spot checking during manufacture. Include a grounding wire in all circuits.

Livestock housing, manure storages, well pits, silos, and silo rooms require special materials and wiring methods, because high levels of moisture and corrosive dust and gas can quickly corrode standard electrical equipment. Dust and moisture can also lead to fire and safety hazards by creating short circuits or heat buildup in electrical components. All wiring hardware must be dust- and moisture-tight and made of corrosion-resistant materials.

Surface mount all wiring on the interior of the building to minimize condensation in wiring system components. Use either a surface-mounted cable or nonmetallic conduit system. Be sure to use appropriate conductors and fittings. Check with your insurance company for their recommendation.

Surface mount cable

Use type UF cable with a grounding wire. Mount cable on the inside surface of walls or ceiling with plastic or plastic-coated staples or straps every 2' and within 8" of junction or fixture boxes. Install cable where it cannot be easily damaged. Install cable on flat surfaces and avoid sharp bends—minimum radius is five times the cable diameter. To cross joints or ribbed metal liners, install the cable on 1x4 running boards.

Use dust- and moisture-tight cable-to-box connectors with rubber or neoprene compression gaskets. Use junction and switch boxes that are drilled and tapped for moisture-tight connectors. **Caution:** do not extend Type UF cable directly into lighting and other heat producing fixtures. High temperatures can damage the insulation on Type UF cable.

Plastic conduit

The most common nonmetallic conduit is Schedule 40 PVC (polyvinyl chloride). Nonmetallic fittings, junction boxes, and conduit are available. PVC conduit is available in 10' and 20' lengths; the most common diameters are ½" and ¾". Conductors designed for use in a wet environment are required. Seal conduit where it passes from a warm to a cold area. Do not extend conduit or cables into the attic space. Use Type THWN or THHN conductors in the conduit.

Switches, outlets, and junction boxes

House every wire splice, switch, and receptacle in a box and mount every fixture on a box. All boxes are to be non-corrosive and dust- and moisture-tight. Use molded plastic or cast aluminum boxes with covers and gaskets. Use receptacle boxes with gasketed, spring-loaded covers. Use switches with either a spring-loaded cover, moisture-tight switch lever, or moisture-tight flexible cover. Bakelite (brown plastic) fixtures are not permitted in livestock buildings.

Locate duplex convenience outlets (DCOs) at least 4' high. Place outlets higher in livestock housing to prevent damage by animals.

Provide one DCO per pen for equipment such as heat lamps or groomers. In stall areas, install at least one on each wall and no more than 20' apart. Provide a DCO wherever portable equipment can be conveniently used. Also install one near each major entrance. DCOs for radio, clock, and desk lamp may be desirable for the office.

Electric Motors

Use totally enclosed farm duty rated motors in environmentally controlled buildings, feed processing rooms, and on equipment subject to dust and moisture accumulation. Use totally enclosed, air-over motors for ventilating fans—the air stream cools the fan.

Provide overload protection for motors in addition to the circuit breakers or fuses. Also, install a fused switch with a time delay fuse at each fan sized at 125%-150% of the motor full load current. Wire each ventilating fan on a separate circuit so that if one circuit fails, fans on another circuit will come on. See *The National Electrical Code Handbook* and *Agricultural Wiring Handbook* for specific wiring requirements.

Service Entrances

If possible, place the service entrance in a dry, clean office or building annex. Exterior-mounted panels with watertight enclosures can be used. If the service entrance is in an animal room, use a weatherproof molded plastic enclosure and mount on asbestos board. Leave a 1″ air space between the service entrance and exterior wall. Install all cable entries at or near the bottom of the service entrance box to prevent condensation from running down cables into the box. Seal cable entries to keep out moisture. Never recess the service entrance box into an outside wall, because water vapor can condense on electrical components, accelerating corrosion.

Determine the required service entrance size based on present needs. Then increase it by 50% to allow for more electrical equipment later. Provide a minimum service entrance capacity of 100 amp, 240 volt. Consider voltage requirements, conductor length, and conductor type when sizing all service conductors. Design for a 2% voltage drop. Refer to MWPS-28, *Farm Buildings Wiring Handbook* and the *Agricultural Wiring Handbook* for tables helpful in sizing service conductors. Contact your electrical power supplier before beginning construction.

Standby Power

A standby power system requires:
• A generator to produce alternating current.
• A stationary engine or tractor PTO to run the generator.
• A transfer switch to isolate the farm system from power lines.

Select either a full- or a partial-load system. A full-load system must handle the maximum running load and peak starting load. A partial-load system carries only enough load to handle vital needs and is usually operated manually. This might include the electrical loads for feed processing and mechanical feeding equipment.

Tractor-driven generators are common and are generally the most economical. During snowstorms, getting the tractor to the generator can be a problem. If short duration outages are critical, consider an engine-driven unit with automatic switching. Operate engine-driven units for short periods at least once a week and tractor-driven units briefly every month to ensure that they will function when needed.

Install a double throw transfer switch at the main service entrance just after the meter, so the generator is always isolated from incoming power lines. This keeps generated power from feeding back over the supply lines, eliminates generator damage when power is restored, and protects power line repair crews. Contact your power suppliers for assistance in sizing and installing a transfer switch.

Sizing a standby power unit is difficult. For a partial-load system, sum the starting wattage of the largest motor, running wattage of all other motors, name plate wattage of essential equipment, and wattage of essential lights. When sizing standby electric generators, consider the power requirements of future expansion and equipment needs.

For more information about standby generators, contact your local power supplier.

Lighting

Three types of lights are common:
• Incandescent.
• Fluorescent.
• High intensity discharge (HID).
Each type has individual properties of light output, maintenance, color, efficiency, and cost that affect selection for a particular task.

Incandescent

Consider incandescent fixtures when light is needed for short periods and when they are turned on and off frequently. Their initial cost is relatively low and they operate well in most conditions, including low temperatures. The efficiency of incandescent lights is relatively low, so they are the most expensive to operate. Use dust- and moisture-tight fixtures with a heat-resistant globe to cover the bulb.

Fluorescent

Fluorescent fixtures cost more than incandescent but produce 3 to 4 times more light per watt. For long on-periods, fluorescent lights are better because of their high efficiency.

Fluorescent lights are used mainly indoors because they are temperature sensitive. Standard indoor lights perform well down to 50 F; with special ballasts, down to −20 F. Humid air above 65% relative humidity makes fluorescent tubes difficult to start.

Use plastic or plastic-coated enclosures with gasketed covers on all fluorescent lights. Select fixtures designed for mildly corrosive industrial environments and equipped with cold start ballasts.

High Intensity Discharge (HID)

HID lamps include mercury, metal halide, high pressure sodium, and low pressure sodium. They tend to have long lives and are energy efficient. They operate well in cold temperatures. Light output is colored, i.e. mercury is greenish-blue and sodium is golden yellow.

HID lamps require 5 to 15 min to start and are not usable where lamps are turned on and off frequently. HID lamps are best when mounted at least 12' high and are on for at least 3 hr. Common uses include feedlots and outdoor security lighting.

Light Levels

Use enough lighting for inspection and to work efficiently, Table 10-1. Provide at least two lighting circuits in each building. Consider two rows of lights on separate switches so two light intensity levels are possible over pen areas.

Install enough switches so at least one light in each room is controlled from main access doors. Consider a separate switch for each row of lights and for lights over all-night feeding stations. Use switches with pilot lights to indicate when outdoor, haymow, and other remote lights are on.

Guidelines for Locating Lights

Provide one light fixture per bull or calving pen. Install additional lighting in pens or handling areas, especially for veterinary care. Provide one row of lights over feedbunks. Add lights to aid specific tasks, brighten dark corners, and avoid long dark shadows. White walls and ceilings improve light levels and are especially useful in areas used to inspect animals. As a rule, space incandescent light fixtures at two times their mounting height along handling and feed alleys.

Feedlot Lighting

Benefits of feedlot lighting are:
• Less trouble with predators and cattle theft.

• Increased animal safety from the quieting effect of night lighting.
• Cattle eat during cool summer nights.
• Reduced stress on newly arrived cattle agitated by darkness.
• Better feed availability for timid cattle.
• Reduced feedbunk space per head, because of 24-hr feeding period (if feed is available).

Lighting to an average of 1 foot-candle in a 30'-50' strip along the feedbunks is recommended. Suspend the lights over the center of a feed alley between 2 rows of bunks. Automatically control the lights to come on at dusk and go off at dawn with a photo cell. For confinement barns, put 150 watt bulbs 20' o.c. over the bunk area. In open lots, high pressure sodium light sources are economical. With high pressure sodium lamps, space poles 225' apart along the bunk. Mount luminaires on 5' arms 35' high. Mercury vapor light sources are also adequate for area lighting but are somewhat more expensive.

Provide light in the corral or handling area, especially for veterinary care. Avoid locating lights directly over feeders and waterers because insects and bird droppings can be a problem.

Lightning

Install lightning arrestors at electrical service entrances to control voltage surges on electrical wiring. Arrestors are particularly important for electronic equipment such as equipment with microprocessor controls and some feed weighing equipment.

Lightning protection for buildings is usually a system of 10"-24" air terminals (lightning rods) with metal conductors attached to a ground. Install lightning rods 20' o.c. and within 2' of the ends of a gable roof. Fasten the metal conductors to roof and walls at

Table 10-1. Light levels for beef buildings.
Table values assume frequent cleaning of bulbs (once a month in animal buildings) and regular bulb replacement (at rated life). With infrequent cleaning and delayed replacement, increase watt/ft² values in the table by 40%.

Design values based on ASAE *Farm Lighting Design Guide,* SP-0175. Assume 8' high ceilings and 70% ceiling and 50% wall reflectance. Incandescent: maintenance factor (MF) = 0.75 and coefficient of utilization (CU) = 0.69; fluorescent: MF = 0.70 and CU = 0.66.

Task	Light level	Standard cool white fluorescent 40 W	Standard incandescent 100 W	150 W
	foot-candles	watt/ft² of building area		
Housing	5	0.14	0.58	0.50
Calving barn	20	0.55	2.29	2.00
Loading platform	15	0.42	1.72	1.50
Animal inspection/handling	20	0.55	2.29	2.00
Along feedbunk	10	0.28	1.15	1.00
Feed storage/processing	10	0.28	1.15	1.00
Haymow	3	0.09	0.35	0.30
Office	50	1.38	5.72	5.00
Toilet room	30	0.83	3.43	3.00

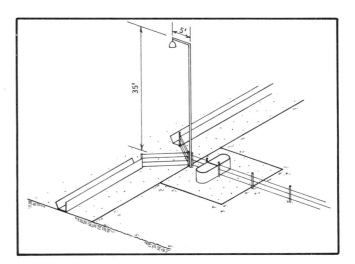

Fig 10-1. Feedlot lighting.
Install sodium vapor lights about 225' apart. Use closer spacing for lower wattage incandescent or mercury vapor lights.

3'-4' intervals. Drive the ground rod at least 8' into moist earth.

Other considerations for barn lightning protection are:
- Ground attached wire fences.
- Extend lightning protection system to an addition.
- Run secondary conductors to mechanical bunks, metal door, etc.
- Install terminals on cupolas, ventilators, etc.
- Provide at least two grounds for the barn.
- Install arrestors on overhead wires.
- Provide one terminal on each domed silo and at least two on each flat-roofed or unroofed silo.
- Ground silo (if required).
- Connect system to water pipes.
- Protect all buildings within 50' of the barn.

Metal clad buildings that have a continuous connection between roofing and siding can be partially protected by grounding the siding unless there is flammable insulation directly beneath the roofing. However, most insurance companies require lightning rods on metal clad buildings as well. Install at least two grounding cables (on opposite corners) on metal clad buildings up to 250' long—add another cable for each additional 100' length or fraction thereof. Make sure your lightning protection installer has qualified for Underwriters' Laboratories "Master Label" designation.

11. MANURE MANAGEMENT

A manure management plan is needed for all parts of a beef production operation. Consider the housing system when selecting a manure management system. Goals of a complete manure management system are to:
- Maintain good animal health through sanitary facilities.
- Minimize air and water pollution.
- Minimize impact on family living areas.
- Reduce odors and dust.
- Control insect reproduction.
- Comply with local, state, and federal regulations.
- Balance capital investment, cash flow requirements, labor, and nutrient use.
- Improve feed efficiency by reducing mud and manure in lots.

Manure can be handled as a solid, semi-solid, slurry, or liquid. The amount of bedding or dilution water influences the form. In turn, the form influences collection and spreading equipment and storage selection.

Solid manure is a combination of urine, bedding, and feces with no extra water added—as found in loafing barns, calving pens, and open lots with good drainage. Semi-solid manure has little bedding and no extra liquid added; little drying occurs before handling. Solid and semi-solid manure can be handled with tractor scrapers, front-end loaders, or mechanical scrapers. Conventional spreaders are common for land application.

Slurry has enough water added to form a mixture capable of being handled by solids handling pumps. Liquid manure is usually less than 8% solids. Large quantities of runoff and precipitation are added to dilute the manure. Liquid and slurry manure is handled with scrapers, flushing systems, gravity flow gutters, or storage below slotted floors. Liquids are spread on fields with tank wagons or by irrigation.

For more information about system selection, design, and management, get MWPS-18, *Livestock Waste Facilities Handbook,* from one of the addresses listed inside the front cover of this book. Get additional help from the Cooperative Extension Service, SCS, consulting engineers, or equipment dealers.

Manure Production

Beef cattle produce about 1.0 ft³ (60 lb) of manure per day per 1,000 lb of liveweight. Fresh manure, a mixture of urine and feces, is about 12% solids, wet basis. Table 11-1 gives weight, volume, and nutrients of manure. Add the weight of bedding for total weight; add ½ the volume of bedding for total volume. Add the weight or volume of water added from leaking waterers, wash water, or any dilution.

Collection

Several collection methods are possible. Consider:
- Facility type.
- Labor requirements.

Table 11-1. Beef manure storage allowance.
Manure at 88.4% water and 60 lb/ft³. No liquids or bedding added. To convert elemental P to P_2O_5 and elemental K to K_2O, divide by 0.44 and 0.83, respectively.

| Animal size, lb | Total manure storage allowance | | | Nutrient content of fresh manure | | |
| | lb/day | ft³/day | gal/day | N | P | K |
				lb/day		
500	30	0.50	3.8	0.17	0.056	0.12
750	45	0.75	5.6	0.26	0.084	0.19
1,000	60	1.0	7.5	0.34	0.11	0.24
1,250	75	1.2	9.4	0.43	0.14	0.31
Cow	63	1.0	7.9	0.36	0.12	0.26

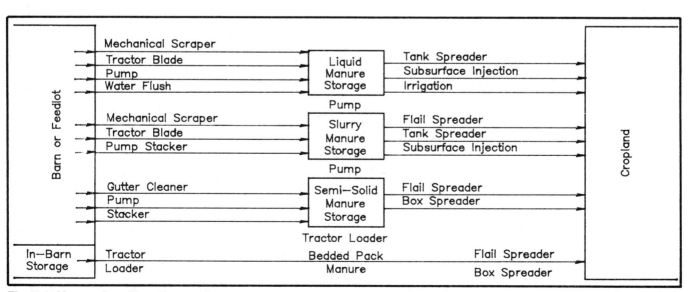

Fig 11-1. Manure handling alternatives.

• Investment.
• Total manure handling system.

Open Lots

Open lots require two manure handling methods. Lot scrapings are solid or semi-solid, and lot runoff is liquid. Move solid manure from the lot to storage with a tractor scraper and front-end loader.

Lot runoff, whether rain or snowmelt, contains manure, soil, chemicals, and debris and must be collected as part of the manure system. Runoff from roofs, drives (but not animal alleys), and grassed or cropped areas without livestock manure, is relatively clean. Divert clean runoff either away from the manure handling system to reduce the total volume or into the system for dilution. Divert clean water away from manure areas with curbs, dikes, culvert pipes, and terraces.

Sheltered Systems

Manure storage can be in a concrete tank under a slotted floor. With outdoor storages, remove manure from the building to storage with a tractor scraper, flushing gutter, or gravity.

Slotted floors

Concrete slats are the most durable and are for animals of all ages. Concrete quality can affect slat durability—see the concrete section of the building construction chapter. Both conventional reinforced and prestressed slats are available. Concrete slats can be precast, precast at the building site, or cast-in-place. Most builders reduce installation labor with precast gang slats.

Slats with a textured surface and tapered sides improve cleaning and provide better footing and wear. Round the edges of homemade slats to prevent chipping, improve cleaning, and reduce foot injuries. Allow for ¼"-½" clearance at each end of the slats for ease of installation. Provide a 1¾"-2" clear opening between slats adjacent to curbs to reduce solids buildup.

Front-end loaders

Typical tractor loaders are available in 1,000 to 4,000 lb sizes. Their turning radius is relatively large and a loaded bucket reduces rear wheel traction, so plan for straight runs, few turns, and nearly level areas. Wide front tractor wheel spacing is desirable for stability. Avoid layouts requiring backing down long alleys.

Skid-steer loaders can clean in cramped areas, greatly reducing hand labor. Most have a turning radius of their own length and a low height, so they can work easily in tight quarters. Some lift relatively low loads.

Flushing systems

In a flushing system, a large water volume carries manure down a sloped gutter to storage. In beef units, flushing under slotted floors, through 3'-4' wide open gutters, and in flumes spaced 12'-14' apart has been used.

Table 11-2. Slat and slot sizes.

Animal type	Slot width in.	Concrete slat width, in.
Cows	1½-1¾	4-8
Feeder cattle	1½-1¾	4-8
Calves	1¼	4-8

Table 11-3. Concrete slat designs.
For more information on slat design, get TR-3, *Concrete Manure Tank Design.*

Length	Depth	Bar
6'	6"	#4
8'	6"	#6
10'	7"	#7

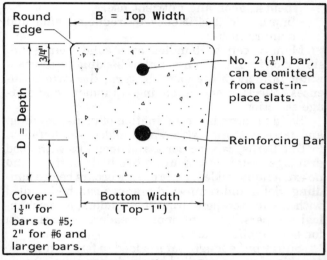

Fig 11-2. Slat cross section.

Install a flush tank or large capacity pump. Use a small pump to fill a flush tank at the high end of the gutter. Flushing frequency can be set by the pumping rate. Make the tank volume at least 75 gal/ft of gutter width and release that volume in about 10 sec. Typically, 100 gal/1,000 lb liveweight per day is needed for adequate cleaning. Determine the flushes per day by dividing the total volume by the volume per flush.

With a large capacity pump, a time clock controls flushing. Pump flushing requires more water. Size the pump for 110 gpm/ft of gutter width. Pump flush for 3 to 5 min at least twice a day.

Flushing flumes are more difficult to maintain than underslat flushing systems. Chipping or scaling of the slot openings results in enlarged openings that can catch an animal's foot and cause injury. Slope gutters 0.5%-1.0%.

A sump and pump at the discharge end of the gutter moves flushed material to a lagoon or earth basin. For gravity transfer, provide at least a 1% slope to the pipe or channel. To reduce the amount of fresh water needed for flushing, recirculate water from the second stage of a lagoon. The flushed materials enter the first cell and the overflow from the second cell is pumped for flushing. Design the inlet of the recycle pump to filter floating scum or trash. Salt and mineral concentrations can plug pumps and distribution pipes. Acetic acid can dissolve these crystals, but land application of lagoon effluent and flushing with fresh water is the best alternative.

Gravity transfer to storage

Gravity transfer of beef manure is being tried. It has worked under slats in beef facilities. Large diameter (20″-36″) pipes work for manure with up to about 3 lb/day of well mixed, chopped, or fine bedding per 1,000 lb animal.

Small diameter (6″-8″) pipes work well with liquid manure. Dry manure or manure with excess bedding may require additional water.

Slope concrete, steel, or plastic pipe at least 5%. Maintain an elevation difference between the collection channel and the top of the storage of 4'-6', Fig 11-3. Bottom load the storage to prevent gases from returning to the building and winter freeze-up. Avoid bends in the pipeline; provide a cleanout at each required bend.

Storage

Evaluate site and soil conditions carefully to avoid water contamination and impact on family living. Avoid locating unlined storages over shallow creviced bedrock, below the water table, in sandy soils or gravel beds, or other areas where serious leakage can cause ground water pollution. Depending on local pollution control regulations, keep the bottom of the storage at least 3' above bedrock and at least 2' above the water table. Allow at least 100' between a water supply and the nearest part of a storage.

Provide for convenient filling, emptying, and surface runoff control. Provide all-weather access and a firm base for the tanker loading area. Design for drive-through loading to avoid backing or maneuvering the tank wagon.

Storage capacity depends on regulations, number and size of animals, amount of dilution by spilled and cleaning water, amount of stored runoff, and desired length of time between emptying. Provide enough storage to spread manure when field, weather, and local regulations permit.

Liquid Manure

Plan for 10 to 12 mo storage capacity. See Table 11-1. Provide extra capacity for dilution water, rain, and snow. If the storage receives only animal manure, add dilution volumes of up to 10% for waterer wastage, rain, and snow. Provide at least 1' of freeboard. From 20%-100% of the storage volume may be needed for dilution water for manure to be irrigated.

Below-ground storage

Storage depth can be limited by soil mantle depth over bedrock, water table elevation, and effective pump lift.

Tanks must be designed to withstand earth, hydrostatic, and live loads, plus uplift if a high water table exists. Columns and beams to support a floor or

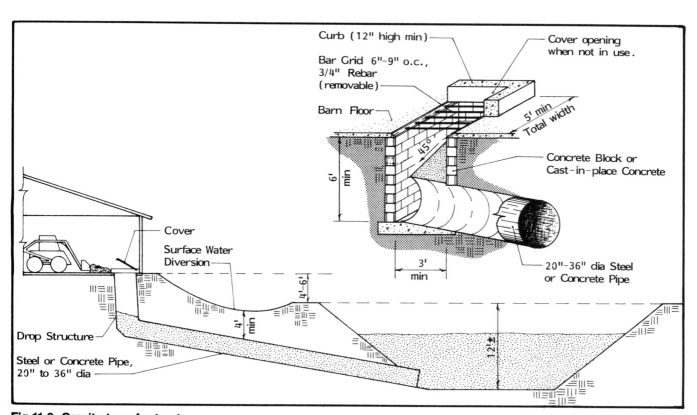

Fig 11-3. Gravity transfer to storage.
Scrape buildings often to prevent drying and maintain flow.

roof are usually 8'-12' apart. For information about concrete tank design, order TR-4, *Welded Wire Fabric in Concrete Manure Tanks*; TR-9, *Circular Concrete Manure Tanks*; TR-10, *Beams for Open-Top Manure Tanks*; and plan mwps-74303, Liquid Manure Tanks—Rectangular, Below Grade, from the addresses listed on the inside front cover of this book.

In cold climates, insulate the upper 2'-3' of exterior tank walls with waterproof insulation. When construction is completed, clean out foreign material that could damage pumps. Before filling with manure, add 6"-12" of water to keep manure solids submerged and counteract the uplifting forces caused by external pressures.

Earth storage basins

Earth basins are at or below grade and may or may not be lined. They can provide long-term storage at low to moderate cost. They are designed and constructed to prevent ground and surface water contamination. Check with your SCS office for help in evaluating site, dike construction, bottom sealing, and basin wall sideslopes. In general, steeper banks conserve space, reduce rainfall in the storage, and leave less manure on the banks when emptied. Inside bank slopes of 1:2 to 1:3 (rise:run) are common for most soils. Make the outside sideslopes no steeper than 1:3 for easier maintenance. Make the embankment wide enough (at least 12') for mowers and agitation equipment.

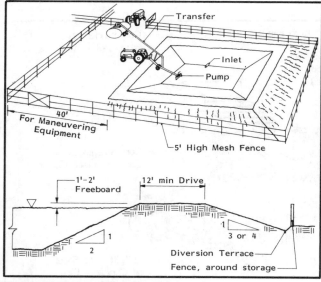

Fig 11-4. Earth storage basins.
Make the berm at least 12' wide for access by agitation, pumping, and mowing equipment. Post warning signs on fence.

Above-ground storage

Above-ground circular tanks are expensive compared to earth basins and are usually not used to store runoff or dilute wastes. However, they are a good alternative where basins are limited by space, high ground water, shallow creviced bedrock, or where earth basins are not aesthetically acceptable.

Above-ground liquid storages are from 10'-60' high and 30'-120' in diameter. They are made of concrete stave, reinforced concrete, and steel. Leaks from joints, seams, or bolt holes can be unsightly, but most small leaks quickly seal with manure. The joint between the foundation and the sidewall can be a problem with improper construction. The reliability of the dealer and construction crew are as important as the tank material in ensuring satisfaction.

Filling

Transfer manure by pump or gravity. With outside storages, locate collection facilities and sumps where manure can be conveniently scraped in. Water may need to be added for easier transfer.

Use bottom loading for beef manure because it will form a surface crust that reduces fly and odor problems. Bottom loading pushes solids away from the inlet and distributes them more evenly. Keep the inlet pipe about 1' above the bottom to prevent blockage.

Agitation

Earth basins are usually agitated with a 3-point hitch or trailer-mounted high capacity manure pump on a long boom. The pump is lowered down the embankment or ramp, placing the pump inlet under the manure surface. The pump chopper and/or rotating auger breaks up the crust and draws solids into the pump for thorough agitation. Provide at least 40' between the berm and fence for moving agitation equipment around the storage.

Agitate above-ground liquid manure tanks by diverting part or all of the pumped liquid through an agitator nozzle. The liquid stream breaks up surface crust, stirs settled solids, and makes a more uniform mixture. Agitate and pump as much manure as possible, then agitate the remaining solids and dilute if necessary. Diluting manure to 90% water may be necessary. Keep surface runoff out of the storage unless included as part of the system design. Use roof runoff for dilution.

Above-ground storage and manure tanks are usually agitated with a centrifugal pump. With silo storages, pumps can be mounted on the storage foundation. Large diameter tanks can have a center agitation nozzle. Agitate below-ground manure tanks with a submerged centrifugal pump, Fig 11-5. Locate agitation sites no more than 40' apart in tanks without partitions (20' o.c. for vacuum pumps). With below-building storage, 10' wide annexes every 40' along the length of the building can be used, Fig 11-6. Locate support columns so they do not interfere with agitation.

Sizing

Determine capacity requirements when planning manure storages. With currently available equipment, about 18"-24" of storage depth is unusable because of agitation requirements and removal efficiency. Determine storage capacity based on a working capacity, which includes manure storage, agitation clearance, precipitation and water storage,

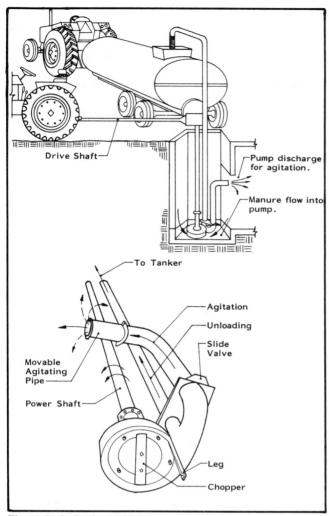

Fig 11-5. Agitating with submerged centrifugal pump.

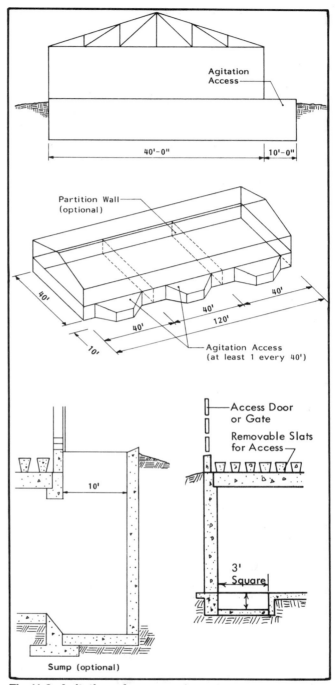

Fig 11-6. Agitation of manure storage.
For wider buildings, provide agitation access on both sides. Get plan mwps-74303, Liquid Manure Tanks—Rectangular, Below Grade, for more details about concrete manure storage construction.

and remaining manure level after emptying. Plan for storing precipitation from a 25-yr, 24-hr storm unless the storage has a roof.

Safety

Protect tank openings with grills and/or covers and enclose open-top tanks and earth basins with a fence at least 5′ high to prevent humans, livestock, or equipment from accidentally entering. Provide a sign alerting visitors about the storage and its hazards. Install railings around pump docks and access points for protection during agitation and cleanout. Provide wheel chocks and tie downs for pumps and tractors.

Gases escaping from agitated manure can be deadly to animals and humans. Operate all ventilating fans and open doors and windows when agitating and unloading manure storages. **Remove animals if possible.**

Never enter a manure storage without a self-contained breathing apparatus and at least two people outside the storage with a rescue line and harness. Do not smoke around a manure storage.

Semi-Solid Manure

Drained storages

Beef manure can be stored and hauled as a solid or semi-solid if ample bedding is used and additional water excluded. Calf pens, maternity pens, and loafing barns are bedded and manure is handled as a solid.

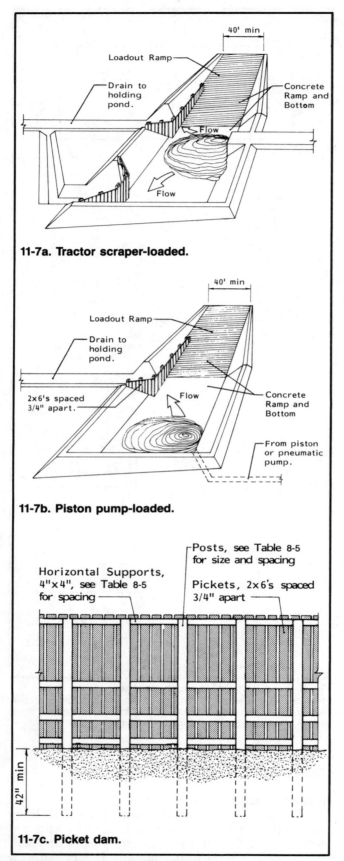

11-7a. Tractor scraper-loaded.

11-7b. Piston pump-loaded.

11-7c. Picket dam.

Fig 11-7. Picket drained storage.
Dam holds manure solids back but allows liquids to drain through.

Table 11-4. Picket dam design.
Posts and horizontal supports are rough sawed timbers. Pickets are 2x6s, rough sawn or surfaced. Use pressure preservative treated wood.

Picket height	Size	Posts Spacing	Horizontal supports Distance from top of pickets	Size	Spacing
0'-4'	4"x6"	5'	0'-4'	4"x4"	3'
5'	6"x6"	4'	4'-6'	4"x4"	2½'
6'	6"x8"	4'	6'-8'	4"x4"	2'
7'	8"x8"	3'			

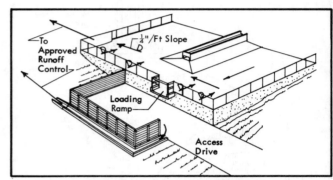

Fig 11-8. Solid manure storage for sloping sites.

A picket dam drains rainwater from manure. This structure has continuous vertical slots about ¾" wide between standing planks or pickets. It holds manure solids back but allows liquids to drain through. Vertical slots work much better than horizontal ones.

Keep all excess water out of the storage. A picket dam only removes rainwater that falls on the storage; it does not reduce the water content of the manure.

The drainage water from the manure storage is polluted; keep it from entering public waters, polluting groundwater, or leaving your property. Direct drainage water to a holding pond or earth basin.

Solid Manure

Up to about 180 days of storage is recommended in cold climates, with less needed in warmer climates. Provide for convenient filling with a tractor mounted manure loader or scraper, elevator stacker, blower stacker, or piston pump. Unload with a tractor mounted bucket. Keep surface runoff out of the storage.

Provide convenient access for unloading and hauling equipment. Slope entrance ramps upward to keep out surface water. Provide a load-out ramp at least 40' wide with a 1:10 slope (1:20 preferred). A roughened ramp improves traction. Angle grooves across the ramp to drain rainwater. Concrete floors and ramps 4" thick are recommended. Slope the floor ⅟16"/ft (½%).

Walls are usually concrete or post-and-plank. Provide one or two sturdy walls to buck against for unloading.

Handling Manure

Solid Manure

Most solid manure spreaders are box-type. Others include flail-type spreaders, dump trucks, earth movers, or wagons. A spreader should distribute manure uniformly. Spreader mechanisms include paddles, flails, and augers.

Liquid Manure

Manure with up to about 4% solids can be handled as a liquid with irrigation or flushing equipment. From 4%-15% solids, manure can be handled as a liquid, but equipment needs differ. Fibrous materials, such as bedding, hair, or feed, can hinder manure pumping. Chopper pumps can cut fibrous materials for improved pumping. Piston manure pumps can handle manure with bedding.

Liquids—up to 4% solids

Liquid manure after solids have been separated or manure with dilution water added can have 4% or less solids. With proper management and screening, a liquid pump is satisfactory, but a slurry or trash pump is more trouble-free. If large quantities are handled, a pipeline may be preferred over tank wagons for transport.

For irrigation, provide an intake screen on the pump, with screen openings no larger than the smallest sprinkler nozzle, spile tube, or gate. A large screen area reduces plugging and velocities into the screen.

Slurries—4% to 15% solids

Solids and liquids separate, so agitate manure before pumping. Pumps moving slurries through long pipelines might operate against fairly high pressures. Pumps for furrow irrigating with gated pipe or for filling tanks usually operate against much lower pressures.

Piston, helical rotor, submerged centrifugal, and positive displacement gear-type pumps can handle heavy slurries against high pressures. However, their performance is improved if solids are below 10%. They do not require priming. To pump heavy slurries against low pressures, use submerged centrifugal, piston, or auger pumps.

Irrigation

Irrigation equipment distributes water and fertilizer on crops. Most irrigation systems can handle liquid manure with up to 4% solids, which is typical of lot runoff and effluent from a lagoon or holding pond. Surface irrigation and big guns can handle higher solids content fluids.

Surface spreading has low cost, low power requirements, and few mechanical parts. Do not use surface irrigation on land with more than a 2% slope.

Sprinklers allow manure application on rolling and irregular land. Although initial and operating costs are generally higher for sprinklers than for surface systems, labor requirements are reduced, systems can be automated, and application uniformity is improved.

Select an irrigation system adapted to your topography, soil, and crops. A well designed and managed system prevents runoff and erosion.

Handling Lot Runoff

Liquid-Solid Separation

Separate solids from liquids by gravity, centrifuges, particle size screens or filters, or water evaporation. After solids are removed, they can be spread as a fertilizer and soil conditioner, used for bedding, or recycled as livestock feed.

Mechanical separators require at least a pump to deliver the manure slurry to the screens and rollers. Power requirements can be up to 25 hp.

Settling basins are gravity separators that remove 50%-85% of the solids from lot runoff. Baffles and porous dams slow the flow enough so settling occurs. A concrete basin floor improves solids removal. Clean out solids as needed between runoff events.

Holding Pond

A holding pond, Fig 11-10, stores runoff water from a settling basin. Size the holding pond for the storage time required in your state (usually 90 to 180 days). A holding pond usually does not receive roof water, cropland drainage, or "clean" water from other sources unless included as part of the design. Base capacity on inches of rainfall on a drained area, such as a 25-yr, 24-hr rainfall event. With additional capacity, emptying the pond can be delayed to fit labor and cropping schedules without fear that another runoff event will cause overflow.

Features of a holding pond:
- The bottom and sides tend to seal naturally. If the pond is in sandy or gravelly soils or near fractured bedrock, seal the bottom by lining with an approved plastic or 6″ (minimum) of compacted clay. Check local requirements.
- It receives runoff, usually from a settling unit, or a lagoon's overflow.
- It is built like an earth storage basin, Fig 11-4.
- Empty the pond by pumping, usually through irrigation equipment, before the storage is full and when liquids will infiltrate into the soil. Avoid spreading on frozen or wet ground.

Vegetative Infiltration Area

Lot runoff treatment by vegetative infiltration areas is practical and economical when grassland or other land that can be kept out of row crop production is available. Limited treatment capacity can be a problem. A settling basin to remove solids is essential. The infiltration area can be a long, 10′-20′ wide channel or a broad, flat area. For uniform distribution, slope the first 50′ 1% to move the runoff rapidly away from the lot and settling unit before further settling occurs. The rest of the area should be nearly flat (less than 0.25%) with just enough drainage to

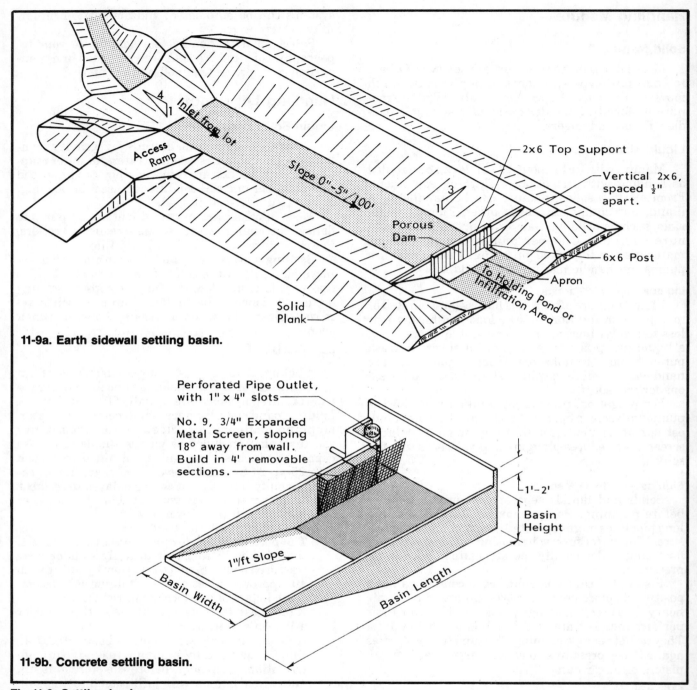

11-9a. Earth sidewall settling basin.

11-9b. Concrete settling basin.

Fig 11-9. Settling basins.

prevent water from standing. The runoff must travel through the vegetated area at least 300' before entering a stream or ditch.

Pumping Manure

Pump Selection

Solids content and required pumping pressure are the major factors in selecting a manure handling pump, Table 11-5. Manure pumps often require re-

pair. Correctly size the pump to reduce breakdowns and need for repair.

Solids content varies with animal age, ration, housing type, and collection system. Settle out solids or add dilution water to reduce pumping problems.

Pressure requirements vary considerably with the intended use and application. A sprinkler irrigation system requires high pumping pressures, but other systems may only need to lift manure 10'-20' to a manure tanker or earth storage.

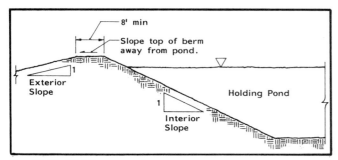

Fig 11-10. Holding pond.
The steepest exterior sideslope is 1:3 (rise:run); 1:4 if it is going to be mowed. Interior sideslope varies from 1:1 to 1:4 depending on the soil construction characteristics. Make the berm at least 12' wide if agitation equipment will travel on top.

Provide an emergency spillway in undisturbed earth near the holding pond dam. Make it 1' below the top of the dam. Overflow from liquid storages is a potential pollutant and must be controlled. Run the spillway overflow to secondary treatment or a vegetative infiltration area. Remember the spillway is for emergency use only.

Land Application and Utilization

Manure application helps build and maintain soil fertility, improve soil tilth, increase water holding capacity, and reduce erosion; runoff or dilute manure adds water.

The fertilizer value of manure spread on land depends on the method of collection, storage, and application, Table 11-6.

Plow down solid manure as soon as possible to minimize nitrogen loss and maximize nutrient utilization. Injecting, chiseling, or knifing liquids into the soil minimizes odors and nutrient losses to the air and runoff. Apply manure as near planting as possible for plant availability. Up to 50% of the total nitrogen can be lost from decomposition, leaching, and runoff when manure is applied in the fall or winter.

Avoid spreading manure on frozen ground where runoff from rain or spring thaws could pollute

Table 11-5. Manure handling pumps.

Pump type	Maximum solids content %	Agitation ability	Agitation range ft	Pumping rate gpm	Pumping head ft of water	Power requirements hp	Applications
Centrifugal							
Open & semi-open impeller Vertical shaft chopper	10-12	Excellent	50-75	1,000-3,000	25-75	65+	Gravity irrigation Tanker filling Pit agitation Transfer to storage
Inclined shaft chopper	10-12	Excellent	75-100	3,000-5,000	30-35	60+	Earth storage agitation Gravity irrigation Tanker filling
Submersible transfer pump	10-12	Fair	25-50	200-1,000	10-30	3-10	Agitation Transfer to storage
Closed impeller	4-6	Fair	50-75	500+	200+	50+	Recirculation Sprinkler irrigation
Elevator	6-8	None	0	500-1,000	10-15	5+	Transfer to storage
Helical screw	4-6	Fair	30-40	200-300	200+	40+	Agitation Sprinkler irrigation Transfer to storage Holding pond and lagoon pumping Tanker filling
Piston							
Hollow	18-20	None	0	100-150	30-40	5-10	Transfer cattle manure without long fibrous bedding
Solid	18-20	None	0	100-150	30-40	5-10	Transfer cattle manure with unchopped bedding
Pneumatic	12-15	None	0	100-150	30-40	—	Transfer to storage
Self loading tanker							
Centrifugal, open impeller	6-8	None	0	200-300	N/A	75+	Tanker loading
Vacuum pump	8-10	Poor	20-25	200-300	N/A	50+	Tanker loading

streams, ponds, open ditches, and groundwater. Excess manure can harm crop growth and yields, contaminate soil, cause surface and groundwater pollution, and waste nutrients. Conduct routine soil tests to monitor soil nutrient levels.

Insect and Rodent Control

Stable flies decrease cattle weight gains and feed efficiency. Build feedbunks and fencelines to minimize manure and debris accumulation that encourages insect breeding and attracts rodents. Set bunks on a solid base, not on cradles or concrete blocks. Put fences on a ridge so manure works away from them. Maintain good lot drainage and proper mound slopes and runoff control facilities to reduce insect breeding.

Remove manure within five days after animals leave to prevent insect larval development. Remove manure, junk, and spilled and spoiled feed from around feed storages, fencelines, bunks, pens, etc. Rodent control requires clean farmsteads.

Insecticides supplement other practices—they do not substitute for good design and management.

Table 11-6. Beef cattle manure nutrients.
Multiply P_2O_5 value by 0.44 to convert to elemental P. Multiply K_2O value by 0.83 to convert to elemental K. One ton of slurry or liquid manure equals about 250 gal.

	Dry matter %	Ammonium N	Total N	P_2O_5	K_2O
		- - lb/ton raw manure - -			
Bedded pack	50	8	21	18	26
Manure pack	50	7	21	14	23
Semi-solid	15	4	11	7	10
Slurry	11	6	10	7	9
Liquid (lagoon)	1	0.5	1	2	1

Table 11-7. Land application of manure.
Table values are in acres/100 animals to yield 100 lb N/acre applied.

Handling method	Feedlot
	acres/100 head
Anaerobic storage	
Broadcast	43.4
Broadcast/cultivate	57.2
Injection	58.8
Irrigation	31.2
Open lot	
Broadcast	21.2
Broadcast/cultivate	32.2
Manure pack	
Broadcast	31.5
Broadcast/cultivate	36.4
Lagoon (irrigated)	13.0

12. FENCES AND GATES

Table 12-1. Fence construction design.

| | Fence | | | | Post | | |
Fence type	Number of strands or rails	Spacing in.	Height to bottom in.	Distance[a] between braces ft	Diameter (top) in.	Height[b] ft	Spacing ft
Plank	4-5	2x6 at 8"	10-14		4	5-6	6-8
Barbed (2-12½ ga)	4-5	8-12	14	1,320	4[e]	4	16-20
High tensile (1-15½ ga)	8	5-8	4-14[c]	60	6	4	16[d]
Cable/strand	4-7	7-12	16-20		5	5	8
Woven or welded wire	—	—	2	660	4[e]	4-5	8
Electric	1-2	16	16-32[c]		3[e]	3	30-50
Suspension	4-6	10	16	1,320	4[e]	4	80-120[f]

[a]Up to ¼ mile on flat terrain.
[b]Add embedment depth (3' min.) to determine post length. Assumes clay loam; sandy or wet soils require greater depths or concrete encasement. For rocky soils where 3' cannot be achieved, use concrete backfill or decrease post spacing.
[c]Lower wire is for calves.
[d]Increase spacing to 60' if battens used.
[e]Steel posts are an option.
[f]Space stays 12'-16'.

Fence construction depends on animal size and numbers, purpose, soil type, topography, and durability required. Temporary fences, such as electric fences for pasture and cropland grazing, are less expensive but also less durable.

Plank or rail fences are strong, economical, easily seen barricades between pens. Plank or rail fences require more maintenance and shading and snow drifting are problems. Barbed and woven wire are for larger pastures where animal crowding, fence pressure, and abuse are less. Welded wire fence panels require reinforcing for use around cattle. Cable fence is for high cattle pressure and abuse; the added expense is offset by lower maintenance.

Posts

Wood

Use pressure preservative treated posts for increased durability. See Table 3-2 for recommended preservative treatment levels. With salt preservative treated lumber, use double hot dipped galvanized or stainless steel nails or other noncorrosive fasteners. Set anchor and line posts at least 3½' in the ground.

Concrete

Concrete posts have minimum upkeep, maximum durability, and are especially resistant to fire. If well made, they last 30 years or longer. Line posts are usually 6" square, Fig 12-1. Several can be made at a time. Shade posts are usually 6"x8" and long enough to provide a shade height of at least 12'. Provide ½" diameter holes for bolting on planks or stringing wire fencing. Be sure holes in posts all face the same direction. Larger anchor and corner posts can be cast in place.

Steel

Steel posts are lightweight, fireproof, durable, and easily driven into most soils. Steel posts also ground a fence against lightning when in contact

Table 12-2. Fence post characteristics.

Post type	Bending strength	Expected life (yrs)	Initial cost	Fire resistance	Maintenance
Steel-T, concrete	Fair	25-30	Medium	Good	Low
Steel rod ⅝" dia.	Poor	15-20	Low	Good	Med
Heavy-duty fiberglass-T	Fair (flexible)	25-30	High	Poor	Low
Light-duty fiberglass-T	Poor (flexible)	15-20	Low	Poor	Med
Pressure treated wood	Good	30-35	Medium	Poor	Very low
Untreated wood	Good	7-15	Low	Poor	High

with wet or moist soil. Common lengths are 5'-7' for line posts for woven wire, electric fence, and barbed wire in firm, well drained soils. Animals crowd steel posts out of line; anchor plates on the posts help.

Plastic and Fiberglass

Plastic and fiberglass posts need no insulators when used with electric fencing. Rot is eliminated but plastic posts have poor fire resistance and are too flexible for corner posts. Use at least 2⅜" diameter round posts and T-posts that are at least 1¼" across. Space wooden line posts every 200' with fiberglass posts 20' apart and every 100' if fiberglass posts are 30' apart.

Battens and Stays

In high tension and some suspension fences, line posts are replaced with less expensive battens or wire stays. Battens are lightweight posts in tension fences and stays are wire spacers in suspension fences. They maintain wire spacing, act as a visual barrier, and help distribute pressures among all wires. With suspension fences, tighten the fence wire so stays barely

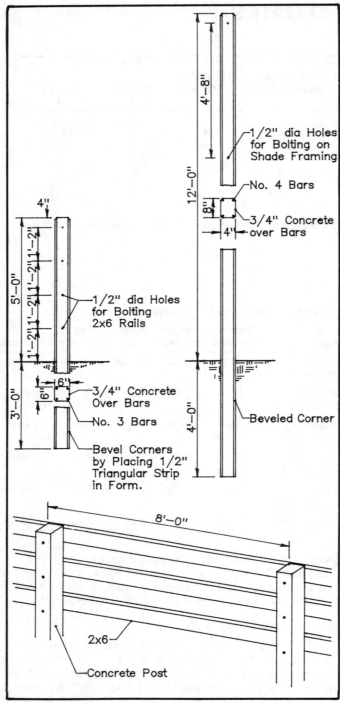

Fig 12-1. Concrete post.
See the building construction chapter for concrete mix recommendations. Use 3,500 psi (minimum) concrete with ½" maximum aggregate.

touch the ground and the fence sways back and forth.

In high tension fences, pressure treated hardwood battens are higher cost, but are stronger and more rigid and easily seen. Angle grooves hold the wire in place without clips; straight grooves require wire clips, Fig 12-2. Drive hardwood battens 18″ into the ground. Hardwood stays can be suspended between posts, but add weight on the wires.

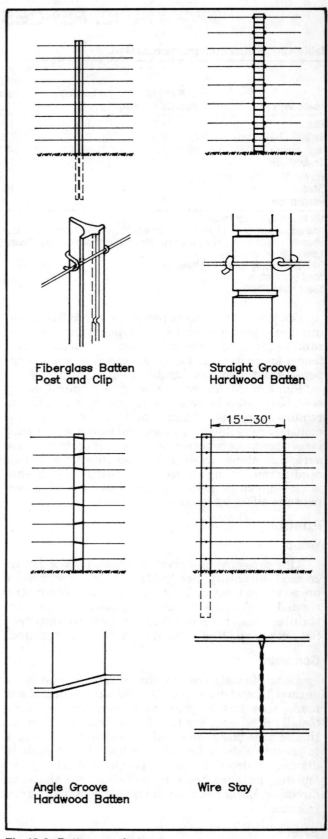

Fiberglass Batten
Post and Clip

Straight Groove
Hardwood Batten

Angle Groove
Hardwood Batten

Wire Stay

Fig 12-2. Battens and stays.
Wood or fiberglass battens for high tension fences. Wire stays are common with suspension fences.

Fiberglass T-posts are light, long lasting, and self insulating. Fiberglass Ts, while not as strong as hardwood, can be set in or suspended above the ground and are usually 2½'-6' long. Use ultraviolet light resistant plastic products.

In suspension fences, stays are commonly twisted wire, light wood slats, or lightweight fiberglass. Lower cost wire stays bend and provide a less effective visual deterrent.

Fencing

High-Tensile

Conventional high-tensile fences have more strands, smooth wire, and tighter wire than conventional barbed wire. The wire is resilient, strong, and comparable in cost, but is stiff, harder to work with, and breaks easily if bent several times. The wire is 9 to 15 gage, galvanized, and has up to 200,000 psi tensile strength and up to 1,800 lb breaking strength. Ordinary mild steel wire has about 25,000 psi tensile strength and 1,000 lb breaking strength.

The splice is usually the weakest part of the fence. Use splices that maintain wire strength, such as a double pressure connector or figure eight knot, Fig 12-3. Mechanical fasteners provide quick and efficient splices and fastenings. If properly installed, fasteners do not slip or weaken the wire. Attach wires to posts with crimping sleeves, knots, or by wrapping and twisting. With high-tensile fencing, in-line stretchers on each wire maintain the proper tension—about 200 to 250 lb in each wire.

Provide high quality corner and brace posts. See equipment section for brace construction. Set corner posts at least 4', and brace posts 3½' deep.

For cattle, a 46" high (minimum), eight wire fence is recommended. On level terrain, space posts every 60' with battens between each post. Use 11 battens between posts for feedlots, 5 battens for pastures, and 3 battens for rangeland.

Suspension

Suspension fences are long spans of barbed wire over level to rolling terrain. Taut (mild or moderate tension) wire that moves freely between fasteners and posts is essential. With wood posts. use 2"-4" long,

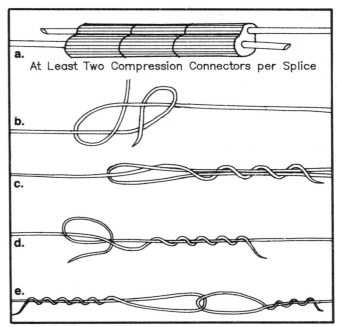

Fig 12-3. Fence splices.

Table 12-3. Splice knot strength.
Strength expressed as a percentage of the breaking strength of wire used. Wire usually breaks at sharp bends. Splice type corresponds to splices in Fig 12-3.

Splice type	Wire type	
	#8 Mild steel	12½ High tensile
	- - - - - Percent - - - - -	
a.	100	100
b.	78	69
c.	76	44
d.	65	28
e.	49	44

9 gage, U-shaped staples; threaded L-shaped staples; or 16 to 20 gage galvanized metal clips. Place line posts every 100' on smooth terrain; reduce post spacing on uneven terrain. Provide stays every 15'-20' between posts. Use double corner braces and single or double-span stretch stations every ¼ mile to maintain fence effectiveness.

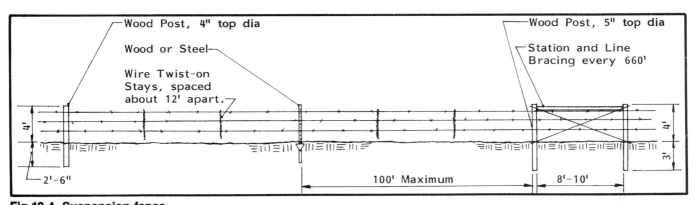

Fig 12-4. Suspension fence.
Staple wire to posts so it moves freely through the staple. Wire movement is important for the fence to be effective.

Cable or Strand Fence

Cable or strand fence works well where cattle are closely confined: feedlots, corrals, etc. Cable fences are strong, durable, and provide unrestricted airflow for better summer cooling, winter thawing, lot drying, and usually fewer snow drifts. A disadvantage is frayed wire causing injuries. Wire rope cables are more flexible and easier to work.

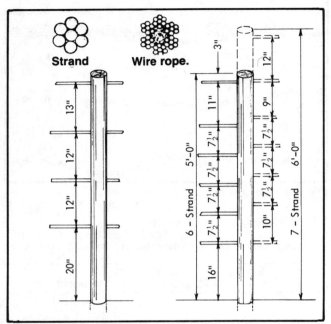

Fig 12-5. Cable fences.
Cable fences are durable, require minimum maintenance, reduce shadows, and improve wind movement. 4″ top posts, 8′ o.c., set 3′-6″ deep (minimum).

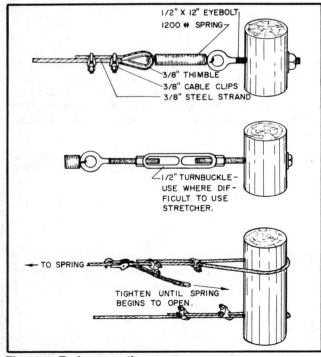

Fig 12-6. End connections.

Use a 4-strand fence for lot partitions and boundaries. Along portions of lot fences that have to restrain driven cattle and near gates and access to working areas, use 6 strands. Use 7-strands for alleys to the working area, and in the working area where there will be no heavy driving, such as holding pen partitions and boundaries.

Table 12-4. Fence post installation.

	Pasture or range	Feedlot fence	Working area
Post length	7′	9′	10′
Min. embedment depth*	3′	3′	3′
Post spacing	16½′	8′	6′

*In muck, peat, or water-saturated soils, increase depth of set about 75% and/or use 6″ or thicker concrete around the post below grade.

Table 12-5. Woven wire fence weights.

	Gage of top and bottom line wires	Gage of filler (intermediate line) wires
Lightweight	11	14½
Medium weight	10	12½
Heavy weight	9	11
Extra heavy wt	9	9

Woven Wire Fences

Most woven wire fencing comes in a choice of weights, protective coatings on the wire, and styles or designs.

Weight

Wire fencing weight depends on the gage of the line or horizontal wires—the lower the gage number the larger the wire and the stronger and more durable the fencing. Field or stock fencing, for example, comes in four weights, Table 12-5.

The stay (vertical) wires are usually the same gage as the filler (intermediate line) wires and may be 6″ or 12″ apart.

Styles

The styles, or designs, of woven fencing are named by a three- or four-digit number, e.g. 932 means 9 line wires in 32″ high fencing, and style 1147, a common cattle fence, has 11 line wires and is 47″ high.

Barbed Wire Fences

Barbed wire is used both alone and with other fencing.

Design

Fig 12-7 shows the usual wire spacing for 2- to 5-strand barbed wire fences.

Barbed suspension fences, Fig 12-4, are common cross and boundary fencing on large cattle ranges. A suspension fence sways back and forth when contacted, striking the cattle, encouraging them to keep away.

Suspension fences have 4 to 6 strands of wire on posts spaced 80′-120′ apart. Twisted wire stays, about 16′ apart, hold the wires apart.

Common range fence is 12½ gage wire with 2-point barbs. In smaller lots where cattle crowd and pressure the fence, 4-point barbs may be more effective. Lighter, 14 gage wire is common for temporary fencing.

High-tensile barbed wire is also available. It is stronger and more durable than comparable sizes of standard wire. For example, 13½ gage high-tensile wire has the breaking strength of 12½ gage standard wire.

Protective coating

Most woven and barbed wire fencing is either zinc (galvanized) or aluminum coated. Zinc coating is Class 1, 2, or 3—the amount of galvanizing/ft² of wire surface, Table 12-6. The thicker the zinc coating, the more corrosion resistance. Class number is usually on a tag on the fencing. Any wire fencing resists corrosion longer in a dry climate than in a humid or industrial one, Table 12-7.

The coating on aluminum-coated fencing is about 0.25 oz/ft² of wire surface and is not usually labeled.

Under the same climatic conditions, aluminum-coated fencing resists corrosion 3 to 5 times longer than zinc-coated with the same coating thickness. But in rural areas, if the coating is broken and the wire exposed to air, zinc protects better. While both metals are "sacrificial agents" (they corrode instead of the wire), in rural atmospheres, an oxide film forms on the aluminum, limiting its ability to protect the wire.

Electric

Electric fences are flexible, low cost, and usually temporary. Temporary fences are one or two electrified wires. More wires, some electrified and others grounded, are more long-term. Maintenance required is moderate. Electric fence success depends on good construction and maintenance.

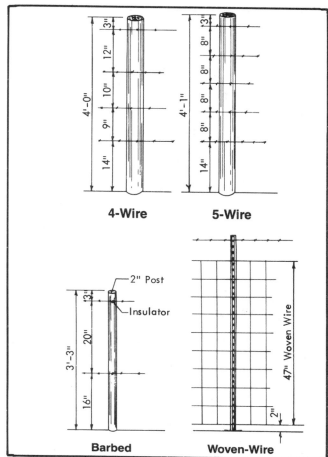

Fig 12-7. Wire fences.

Table 12-6. Zinc protective coatings.

Wire gage	Zinc coating, oz/ft² of wire surface		
	Class 1	Class 2	Class 3
9	0.10	0.60	0.80
10	0.30	0.50	0.80
11	0.30	0.50	0.80
12½	0.30	0.50	0.80
14½	0.20	0.40	0.60

Table 12-7. Wire fence durability.

Wire size	Climatic condition			
	Dry		Humid	
	Class			
	1	3	1	3
	Years till rust appears			
9	15	30	8	13
11	11	30	6	13
12½	11	30	6	13
14½	7	23	5	10

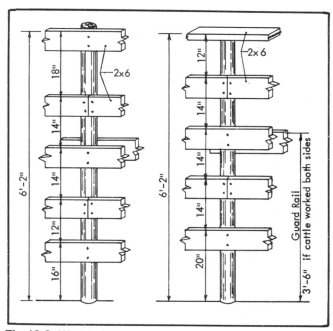

Fig 12-8. Working fences.

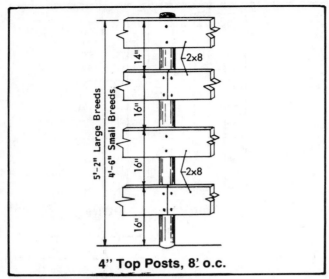

Fig 12-9. Feedlot line fences.
Set line fences on a mound or ridge for drainage, to work manure away, and reduce cattle's tendency to jump.

Temporary fences can be plastic cord with metal strand or smooth No. 12 galvanized wire. Barbed wire is stronger and more visible than lighter gage smooth wire. Although harder to handle, the barbs make better electric contact with the animal.

An electric fence has a controller, wire, sturdy posts, and ground rods. Unless posts are fiberglass, fasten the wire to posts with insulators.

Typical post spacing is 10'-30'. On flat terrain, spacing can be increased. Space posts on hilly terrain to maintain wire height. Use 4" minimum diameter corner posts 36" deep, and thoroughly tamp the backfill even for temporary fences. Double-brace end and corner posts of permanent electric fences.

A single wire electric fence may not work over very dry or frozen soil or snow which acts as an insulator, reducing the effectiveness of the charged wire. Add another wire 6" above the charged wire and connect it to ground rods at convenient intervals along the fence. Animals must contact both wires to get a full shock.

Hang shiny objects on fence wires to train cattle that wire is electrified. Post warning signs to notify children and workers and reduce accidents.

Electric fence controllers

Select a controller listed by Underwriters' Laboratories (UL) or comparable safety inspection agency. Notice of approval is printed on the controller near the name plate. Follow manufacturer's recommendations. Homemade or nonapproved chargers can cause serious injury to humans and animals and are not recommended.

Power the controller with 6 or 12 volt direct current (dc) batteries or a 120 volt alternating current (ac) source. Combination units are available, as are solar collector and wind charged battery units.

Fence controllers are carefully designed to intermittently charge the fence wire. Battery operated

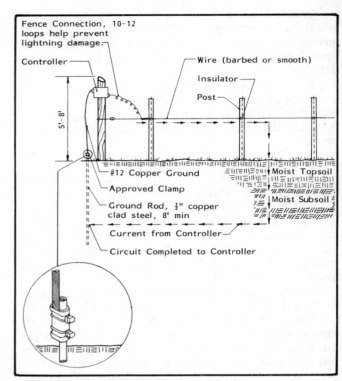

Fig 12-10. Features of an electric fence.
Make good connections between wires to be electrified.

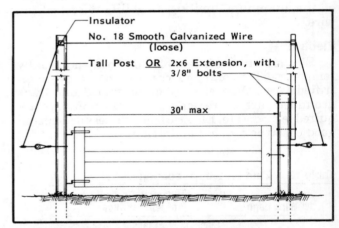

Fig 12-11. Electric fence above a swing gate.

controllers commonly supply 1,000 to 2,500 volts. Alternating current controllers supply about 3,000 to 6,000 volts. The controller develops an "on-off" current cycle so those touching the wire can be freed from the wire. The "on" time is 1/10 of a second or less, 45 to 55 times a minute, which is safe for humans and effective for livestock. The shock is sharp, but short and relatively harmless.

Locate the fence controller inside to be accessible for inspection and to be protected from weather and livestock. If it must be outdoors, weatherproof units are available. If the controller is not weatherproof, enclose it, Fig 12-12, leaving space for the controller, a battery, and for making connections. Seal the enclosure and face it toward the sun to aid drying.

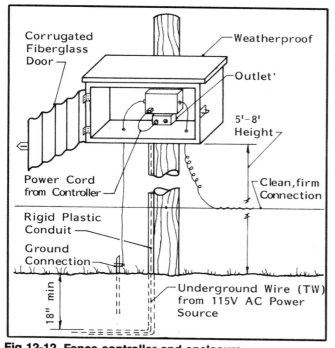

Fig 12-12. Fence controller and enclosure.
Use buried-wire insulation rated for at least 10,000 volts where electric fence wire passes underground.

Proper grounding is essential—it is the return side of the circuit. Improper grounding is a common problem in controller installation. If connections are good, the current travels from the controller to the fence wire, animal, soil, ground rod, and back to the controller.

A ¾″ diameter galvanized pipe, ⅝″ solid steel rod, or ½″ copper clad steel rod driven at least 8′ into continually damp soil provides a low resistance pathway for electricity to ground. Clamp the wire leading from the controller ground terminal securely to the ground rod with an approved ground rod clamp. Do not use a well, water pipe, or power ground system to ground an electric fence.

Insulators

Fence insulators are needed on all but fiberglass posts to prevent grounding the electric wire and making it ineffective. Porcelain is the best material. High density plastic insulators fit various post types but can deteriorate and become brittle in sunlight. Carefully check manufacturer recommendations about voltage rating and installation.

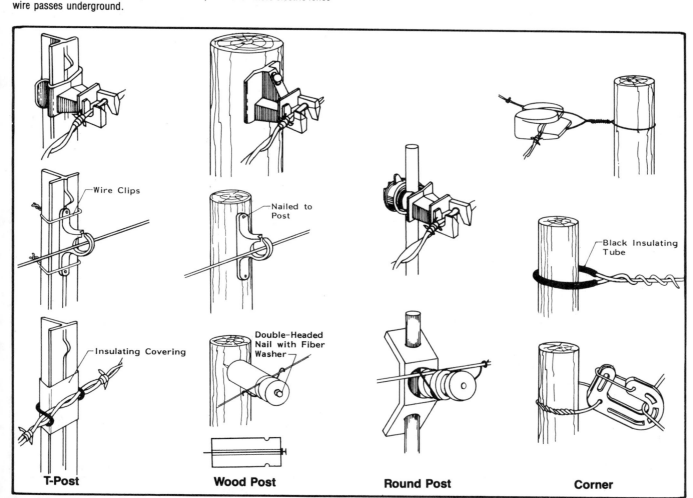

Fig 12-13. Electric fence insulators.

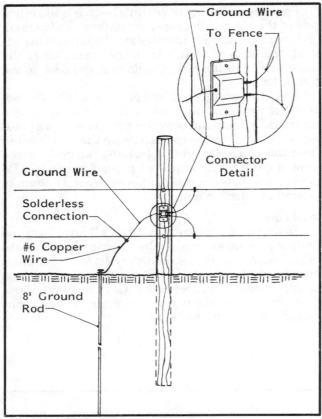

Fig 12-14. Electric fence lightning arrestor.

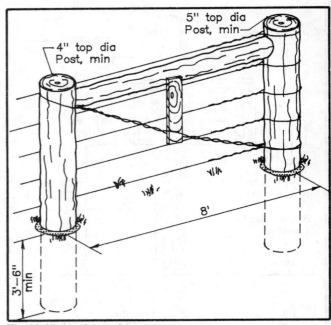

Fig 12-15. Horizontal brace.
Add another post and brace for longer stretches, tighter fences, and lighter soils. See Fig 12-19 for more details.

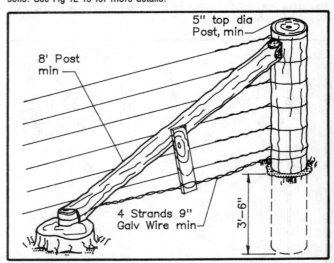

Fig 12-16. Diagonal brace.

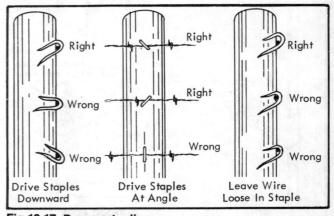

Fig 12-17. Proper stapling.
Deformed shank staples have better holding power, especially in penta and creosote treated posts.

Fence Braces

Braces maintain wire tension and fence effectiveness and can increase fence life. The most common system is the horizontal brace assembly, Fig 12-15. A diagonal brace on one post is also used, Fig 12-16. With diagonal braces, end posts tend to pull upward and out sooner than with horizontal braces. Bracing end and corner posts depends on post diameter, embedment, height, soil type, moisture variation, and freeze-thaw cycles. See Table 12-1 for recommended brace spacing and Fig 12-18 for construction methods.

Staples

Staple length, diameter, smooth or deformed shank, point type, and post type affect holding power. For pressure preservative treated posts, use 1¾", 9 gage, hot dip galvanized staples with slash cut points. Shorter staples with single legs or diamond points have less holding power unless hardened and grooved, especially with creosoted posts.

String wires on the cattle side of posts and on the outside of curves. Drive staples off the vertical, so they straddle the wood grain and wires move freely. Rotate staples 20°-30° from the flat side of the point, Fig 12-17, to spread the legs for greater holding power.

Equipment

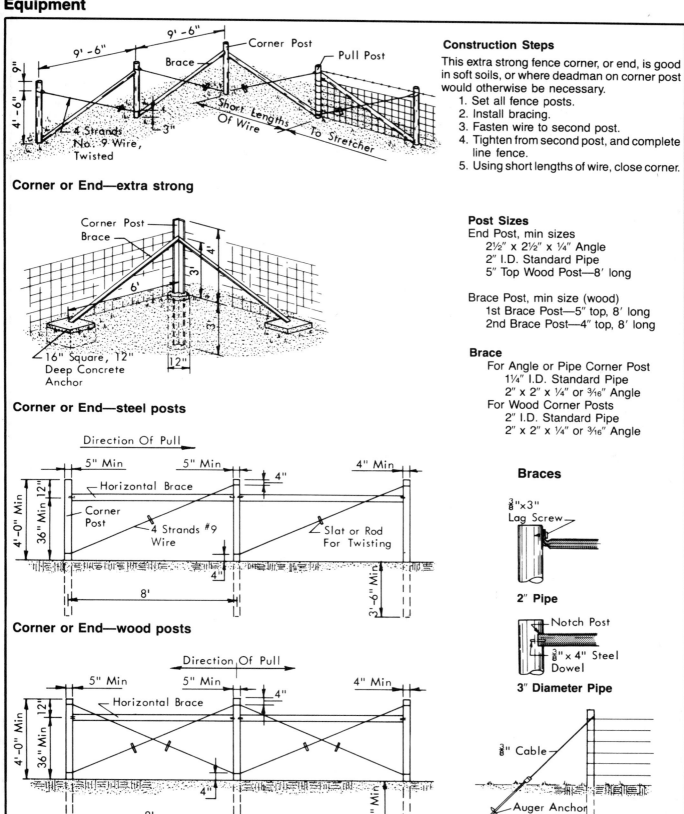

Corner or End—extra strong

Corner or End—steel posts

Corner or End—wood posts

Wood Pull Post
For middle of long fence, place about 660' apart.

Fig 12-18. Fence corner and board brace construction.

Construction Steps

This extra strong fence corner, or end, is good in soft soils, or where deadman on corner post would otherwise be necessary.

1. Set all fence posts.
2. Install bracing.
3. Fasten wire to second post.
4. Tighten from second post, and complete line fence.
5. Using short lengths of wire, close corner.

Post Sizes

End Post, min sizes
2½" x 2½" x ¼" Angle
2" I.D. Standard Pipe
5" Top Wood Post—8' long

Brace Post, min size (wood)
1st Brace Post—5" top, 8' long
2nd Brace Post—4" top, 8' long

Brace

For Angle or Pipe Corner Post
1¼" I.D. Standard Pipe
2" x 2" x ¼" or ³⁄₁₆" Angle
For Wood Corner Posts
2" I.D. Standard Pipe
2" x 2" x ¼" or ³⁄₁₆" Angle

Braces

2" Pipe

3" Diameter Pipe

Auger-anchors can brace corners and ends.

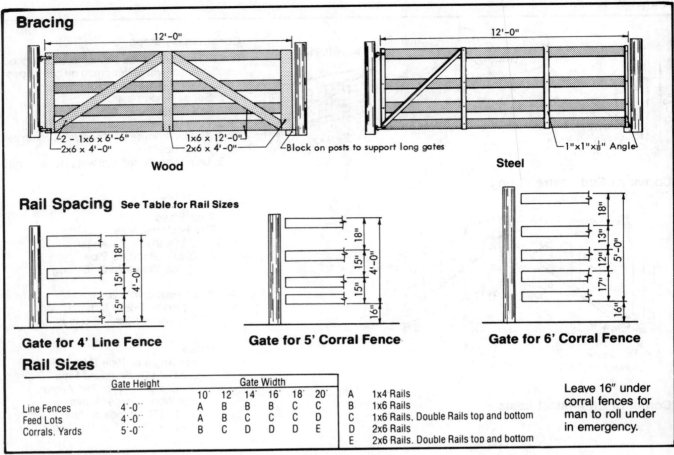

Bracing

Wood

12'-0"

2 - 1x6 x 6'-6"
2x6 x 4'-0"
1x6 x 12'-0"
2x6 x 4'-0"
Block on posts to support long gates

Steel

12'-0"

1"x1"x⅛" Angle

Rail Spacing See Table for Rail Sizes

Gate for 4' Line Fence

18"
15"
15"
4'-0"

Gate for 5' Corral Fence

18"
15"
15"
4'-0"
16"

Gate for 6' Corral Fence

18"
13"
12"
17"
5'-0"
16"

Rail Sizes

	Gate Height	Gate Width					
		10'	12'	14'	16'	18'	20'
Line Fences	4'-0"	A	B	B	B	C	C
Feed Lots	4'-0"	A	B	C	C	C	D
Corrals, Yards	5'-0"	B	C	D	D	D	E

A   1x4 Rails
B   1x6 Rails
C   1x6 Rails, Double Rails top and bottom
D   2x6 Rails
E   2x6 Rails, Double Rails top and bottom

Leave 16" under corral fences for man to roll under in emergency.

Fig 12-19. Wood gates.

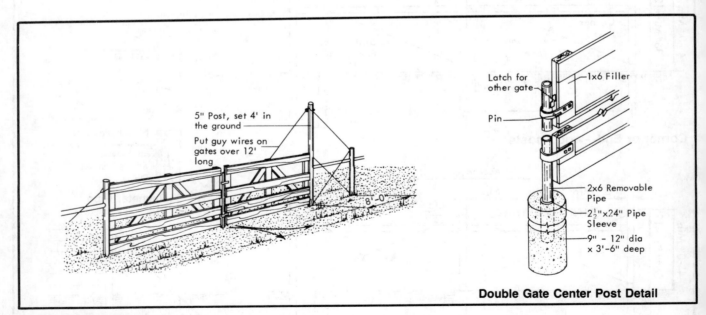

5" Post, set 4' in the ground

Put guy wires on gates over 12' long

8'-0"

Latch for other gate
Pin
1x6 Filler
2x6 Removable Pipe
2½"x24" Pipe Sleeve
9" – 12" dia x 3'-6" deep

Double Gate Center Post Detail

Fig 12-20. Double gate.

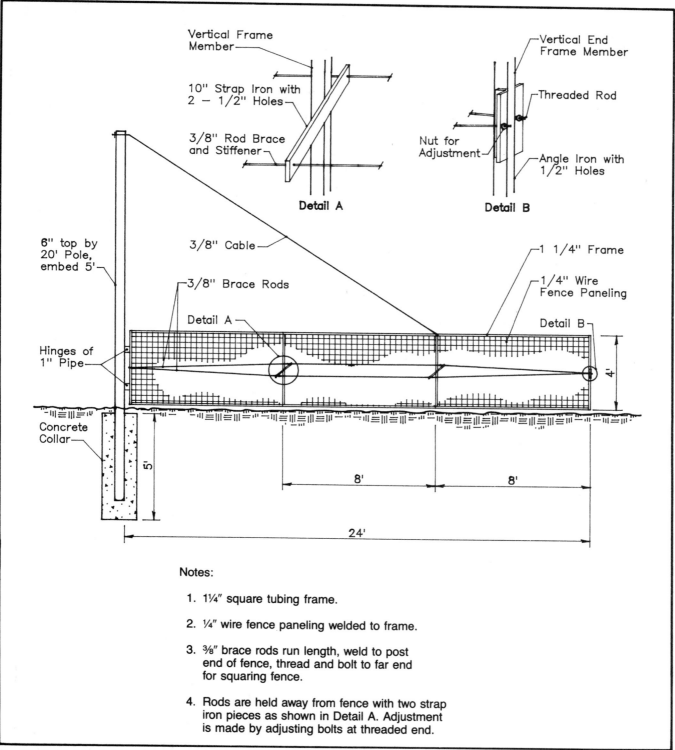

Vertical Frame Member

10" Strap Iron with 2 — 1/2" Holes

3/8" Rod Brace and Stiffener

Detail A

Vertical End Frame Member

Threaded Rod

Nut for Adjustment

Angle Iron with 1/2" Holes

Detail B

6" top by 20' Pole, embed 5'

3/8" Cable

3/8" Brace Rods

Detail A

1 1/4" Frame

1/4" Wire Fence Paneling

Detail B

Hinges of 1" Pipe

Concrete Collar

4'

5'

8'

8'

24'

Notes:

1. 1¼″ square tubing frame.

2. ¼″ wire fence paneling welded to frame.

3. ⅜″ brace rods run length, weld to post end of fence, thread and bolt to far end for squaring fence.

4. Rods are held away from fence with two strap iron pieces as shown in Detail A. Adjustment is made by adjusting bolts at threaded end.

Fig 12-21. Lightweight gate construction.

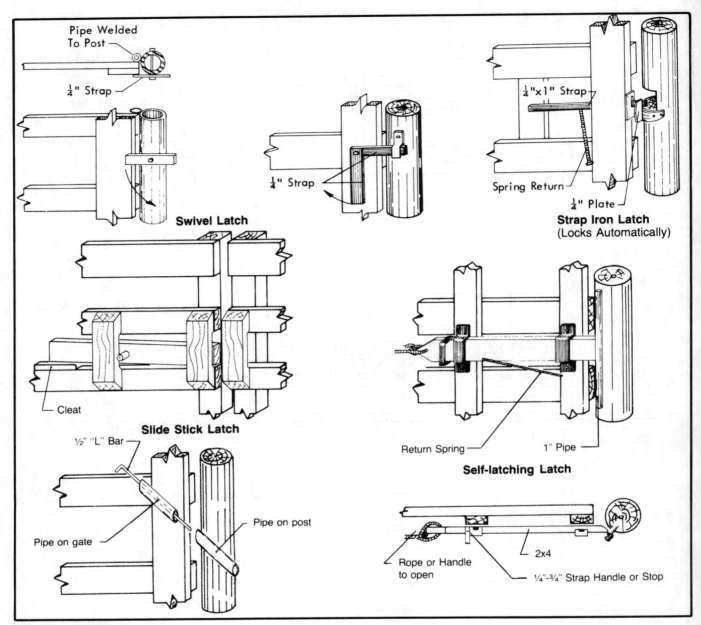

Pipe Welded To Post

¼" Strap

Swivel Latch

¼" Strap

¼"x1" Strap

Spring Return

¼" Plate

Strap Iron Latch
(Locks Automatically)

Cleat

Slide Stick Latch

½" "L" Bar

Pipe on gate

Pipe on post

Return Spring

1" Pipe

Self-latching Latch

Rope or Handle to open

¼"-¾" Strap Handle or Stop

2x4

Fig 12-22. Latches.

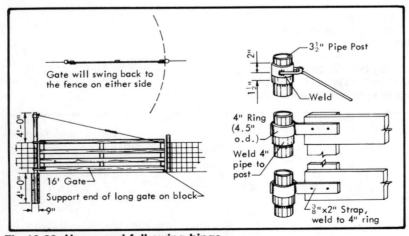

Gate will swing back to the fence on either side

4'-0"

16' Gate

4'-0"

Support end of long gate on block

9"

3½" Pipe Post

2"

1½"

Weld

4" Ring (4.5" o.d.)

Weld 4" pipe to post

⅜"x2" Strap, weld to 4" ring

Fig 12-23. Heavy and full swing hinge.

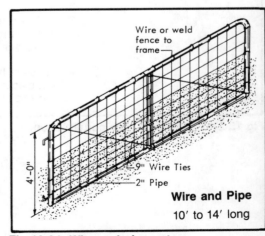

Wire or weld fence to frame

4'-0"

9" Wire Ties

2" Pipe

Wire and Pipe
10' to 14' long

Fig 12-24. Wire and pipe gate.

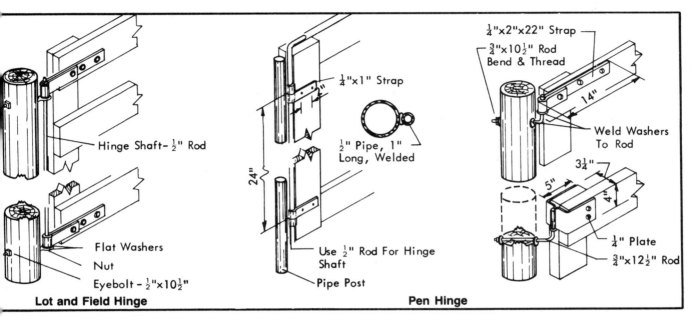

Fig 12-25. Lot and pen hinges.

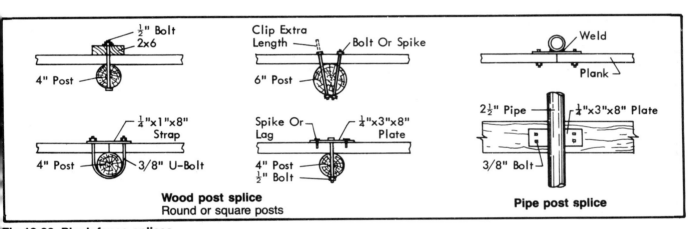

Fig 12-26. Plank fence splices.

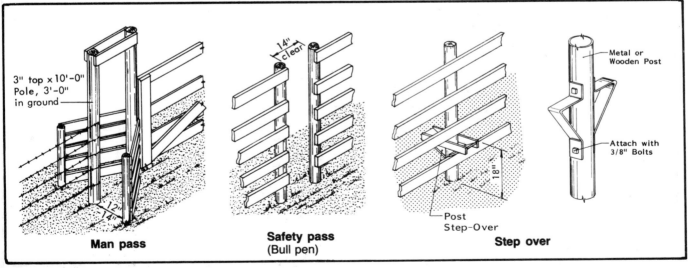

Fig 12-27. Stiles and passes.

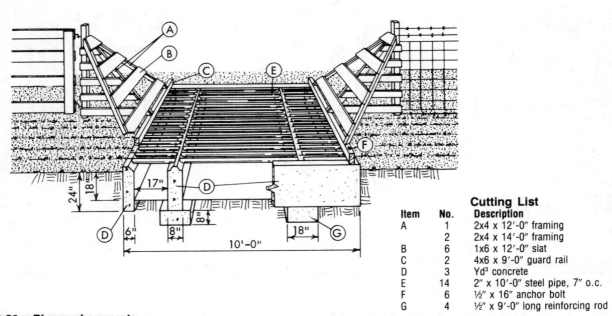

Cutting List

Item	No.	Description
A	1	2x4 x 12'-0" framing
	2	2x4 x 14'-0" framing
B	6	1x6 x 12'-0" slat
C	2	4x6 x 9'-0" guard rail
D	3	Yd³ concrete
E	14	2" x 10'-0" steel pipe, 7" o.c.
F	6	½" x 16" anchor bolt
G	4	½" x 9'-0" long reinforcing rod

12-28a. Pipe and concrete.

Minimum width of guards 5'-0"
Recommended width of guards 8'-0"
Provide gate to one side of guard for passage of cattle and extra heavy trucks, tractors, and wide machinery.
All lumber pressure-treated. If used where cattle are apt to be crowded (feedlots, alleys) provide gate on cattle side.

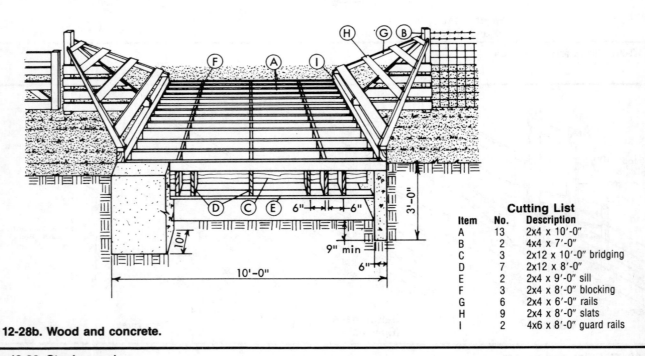

Cutting List

Item	No.	Description
A	13	2x4 x 10'-0"
B	2	4x4 x 7'-0"
C	3	2x12 x 10'-0" bridging
D	7	2x12 x 8'-0"
E	2	2x4 x 9'-0" sill
F	3	2x4 x 8'-0" blocking
G	6	2x4 x 6'-0" rails
H	9	2x4 x 8'-0" slats
I	2	4x6 x 8'-0" guard rails

12-28b. Wood and concrete.

Fig 12-28. Stock guards.

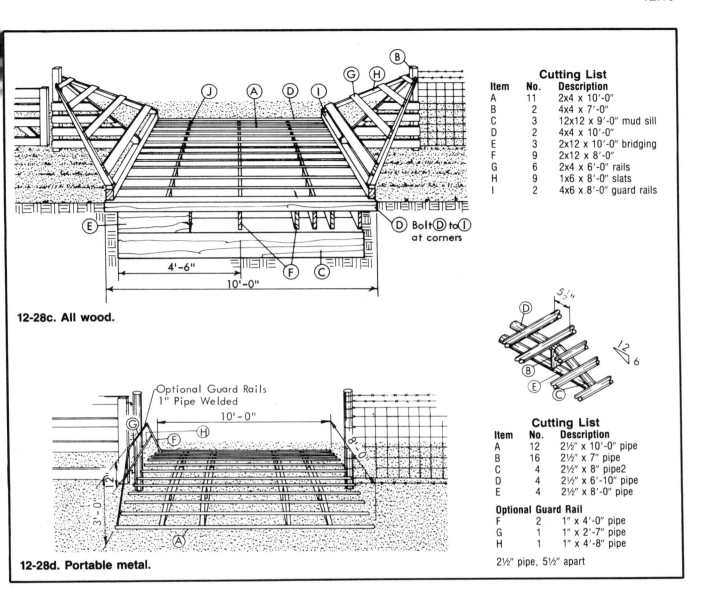

12-28c. All wood.

Bolt Ⓓ to Ⓘ at corners

4'-6"

10'-0"

Cutting List

Item	No.	Description
A	11	2x4 x 10'-0"
B	2	4x4 x 7'-0"
C	3	12x12 x 9'-0" mud sill
D	2	4x4 x 10'-0"
E	3	2x12 x 10'-0" bridging
F	9	2x12 x 8'-0"
G	6	2x4 x 6'-0" rails
H	9	1x6 x 8'-0" slats
I	2	4x6 x 8'-0" guard rails

12-28d. Portable metal.

Optional Guard Rails 1" Pipe Welded

10'-0"

8'-0"

12"

3'-0"

5½"

12 / 6

Cutting List

Item	No.	Description
A	12	2½" x 10'-0" pipe
B	16	2½" x 7" pipe
C	4	2½" x 8" pipe2
D	4	2½" x 6'-10" pipe
E	4	2½" x 8'-0" pipe

Optional Guard Rail

F	2	1" x 4'-0" pipe
G	1	1" x 2'-7" pipe
H	1	1" x 4'-8" pipe

2½" pipe, 5½" apart

13. SELECTED REFERENCES

Available from the Extension Agricultural Engineer at any of the institutions listed on the inside front cover or from Midwest Plan Service.

MWPS-1 *Structure and Environment Handbook.*

MWPS-2 *Farmstead Planning Handbook.*

MWPS-13 *Grain Drying, Handling and Storage Handbook.*

MWPS-14 *Private Water Systems Handbook.*

MWPS-18 *Livestock Waste Facilities Handbook.*

MWPS-28 *Farm Buildings Wiring Handbook.*

MWPS-35 *Farm and Home Concrete Handbook.*

AED-20 *Managing Dry Grain in Storage.*

AED-23 *Outside Liquid Manure Storages.*

TR-4 *Welded Wire Fabric in Concrete Manure Tanks.*

TR-9 *Circular Concrete Manure Tanks.*

TR-10 *Beams For Open-Top Manure Tanks.*

Available from the Extension Agricultural Engineer at any of the institutions listed on the inside front cover, Midwest Plan Service, or Northeast Regional Agricultural Engineering Service (NRAES), Riley-Robb Hall, Cornell University, Ithaca, NY 14853. Ph. 607-256-7654.

NRAES-1 *Pole and Post Buildings.*

NRAES-11 *High-Tensile Wire Fencing.*

NRAES-18 *Extinguishing Silo Fire.*

Other resources.

Agricultural Wiring Handbook. Food and Energy Council, Inc., Columbia, MO 65202. 1978.

Cement Mason's Guide. Portland Cement Association, Skokie, IL 60077-4321. 1980.

Concrete-Paved Feedlots. Portland Cement Association, Minneapolis, MN 55435.

Design and Control of Concrete Mixtures. Portland Cement Association, Skokie, IL 60077-4321. 1979.

Electrical Wiring Systems for Livestock and Poultry Facilities. H. David Currence, National Food and Energy Council, Inc., Columbia, MO 65202. 1983.

Farm Lighting Design Guide, SP-0175. American Society of Agricultural Engineers, St. Joseph, MI 49085.

Fire In Silos, Prevention and Extinguishing. International Silo Association, Inc., West Des Moines, IA 50265.

Livestock Environment, Proceedings of the International Livestock Environment Symposium. American Society of Agricultural Engineers, St. Joseph, MI 49085. 1974.

Livestock Environment II, Proceedings of the second international livestock environment symposium. American Society of Agricultural Engineers, St. Joseph, MI 49085. 1982.

National Design Specification. National Forest Products Association, Washington, DC 20036.

The National Electrical Code Handbook. National Fire Protection Association, Quincy, MA. 1980.

Resurfacing Concrete Floors. Portland Cement Association, Skokie, IL 60077-4321. 1981.

14. SELECTED PLAN REFERENCES

MWPS Beef Plans

Several beef building, hay barn, manure storage, and other plans and publications are available from the Extension Agricultural Engineer at any of the institutions listed on the inside front cover of this handbook or from Midwest Plan Service. Send for a catalog from any of the institutions listed on the inside front cover or from Midwest Plan Service.

USDA Beef Plans

The following USDA beef plans are available from the Extension Agricultural Engineer at any of the institutions listed on the inside front cover of this handbook.

Cattle Holding Chute and Headgate   USDA-5778

Stationary; wood and metal construction; openable side panels; roofed squeeze area.

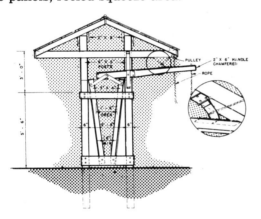

Vat for Dipping Cattle   USDA-5876

Reinforced concrete and wood; slide entry.

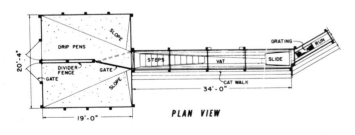

Cattle Dipping Vat and Inspection Facility   USDA-5940

Step/ledge entry; concrete and wood; covered inspection area.

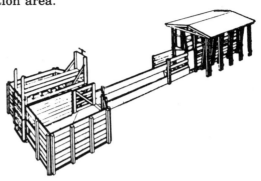

Variable Height Loading Chute   USDA-5852

Permanent; wood; pole type with cleated ramp. See Fig 7-24.

Cutting Gate, Tower and Chute   USDA-5959

With tapered chute and overhead tower/work platform; wood and steel construction.

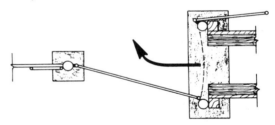

Automatic Crowd Gate   USDA-6037

For circular holding area; cable and weight control yields continual, "automatic" crowding.

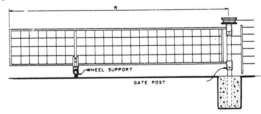

Portable Corral Transport   USDA-6041

Folds to 8'x20' for transport; steel construction.

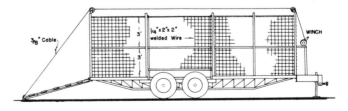

Cattle Feeders USDA-6066

Portable; feed bunk (pipe frame) and supplement feeder.

Breeding Chute for Beef Cattle USDA-6103

Roofed; wood. See Fig 5-11.

Squeeze Chute, Variable Width, Trailer Mounted USDA-6133

Variable width; openable, straight sides; steel; trailer mounted.

Trailer For Cattle, Gooseneck Type USDA-6141

6'x16'; wood and steel.

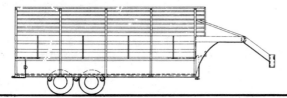

Gate USDA-6151

Telescoping type; all steel.

Corral and Breeding Chute for Beef Cattle USDA-6152

2 triangular pens; wood. See Fig 5-13.

Cable Fencing USDA-6162

Thru-post cables; spring tighteners.

Dock Bumper USDA-6177

Self aligning; steel and wood. See Fig 7-25.

3 Tier Loading Chute USDA-6183

Wood; cleated ramp; includes adjustable ramp section. Similar to Fig 7-24.

Pick-up Truck Racks USDA-6192

Light and heavy duty details shown. 8' long; 6' wide; center opening and side opening rear gate option.

Corral Layout USDA-6205

Semi-circular layout with perimeter alley and pie-shaped pens; central working facilities with semi-circular chute.

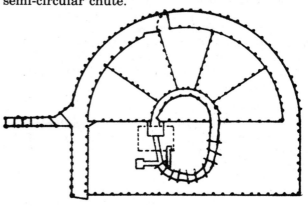

Hay Feeders for Round Bales USDA-6214

Wood and steel; circular and rectangular. Similar to Figs 8-27 and 8-29.

Portable Mineral Feeder USDA-6226

8'x10'; roofed; wood; skid mounted.

Expansible Corrals USDA-6229

Two layouts shown; expandable semi-circular chute; tub crowding area; steel and wood; pie-shape or rectangular pens. See Fig 7-8a.

Corrals With Working Facilities USDA-6230

Two layouts shown; rectangular arrangement; in-line or curved working facilities; tub crowding area; steel and wood; rectangular pens. See Fig 7-8b.

Slant Bar Feeder Panel USDA-6242

Square or round steel tubing. See Fig 8-30.

Covered Feeders for Round Bales USDA-6245

Roofed; steel; skid mounted; 8'x12'.

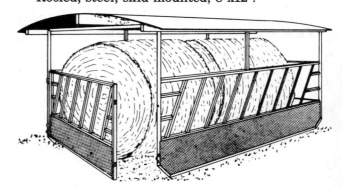

Pasture Creep Gate **USDA-6266**
For calves; steel or wood.

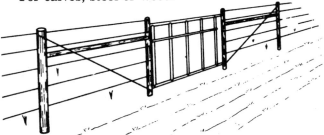

Variable Clearance Gate **USDA-6365**
Details for two opposing gates to permit opening of one large, one small, or both gates; clearance above ground variable from 6"-24".

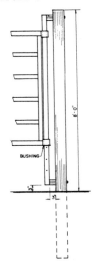

Lightning Protection **USDA-6368**
Buildings; plan shows placement of air terminals for different sizes, types and shapes of buildings plus silos; details shown for bonding and grounding.

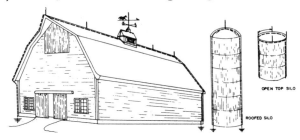

15. INDEX